Structure and Properties of Lengthy Rails after Extreme Long-Term Operation

A.A. Yuriev, V.E. Gromov, Yu.F. Ivanov, Yu.A. Rubannikova,
M.D. Starostenkov, P. Y. Tabakov

Published by **Materials Research Forum LLC**
Millersville, PA 17551, USA

Published as part of the book series
Materials Research Foundations
Volume 106 (2021)
ISSN 2471-8890 (Print)
ISSN 2471-8904 (Online)

Print ISBN 978-1-64490-146-5
ePDF ISBN 978-1-64490-147-2

This book contains information obtained from authentic and highly regarded sources. Reasonable efforts have been made to publish reliable data and information, but the author and publisher cannot assume responsibility for the validity of all materials or the consequences of their use. The authors and publishers have attempted to trace the copyright holders of all material reproduced in this publication and apologize to copyright holders if permission to publish in this form has not been obtained. If any copyright material has not been acknowledged please write and let us know so we may rectify in any future reprint.

Distributed worldwide by

Materials Research Forum LLC
105 Springdale Lane
Millersville, PA 17551
USA
http://www.mrforum.com

Printed in the United States of America
10 9 8 7 6 5 4 3 2 1

Table of Contents

Introduction

Currently, railways in the world account for up to 85% of freight turnover and over 50% of passenger traffic. Recently, there has been a significant increase in the intensity of railway transport and its load intensity, which requires high operational durability of rails. To solve these problems, the technology of differentiated hardening of 100-meter rails is used, the production of which began in Russia in 2013. The processes of formation and evolution of the structural-phase states and properties of the surface layers of rails during long-term operation represent a complex set of interrelated scientific and technical issues. The importance of information in this area is determined by the depth of understanding of the fundamental problems of condensed matter physics, on the one hand, and the practical significance of the problem, on the other [1-3].

Improving the modes of differentiated hardening of long rails for the formation of high operational properties should be based on the knowledge of the mechanisms of structural and phase changes over the cross-section of the rails during their long-term operation. The identification of such mechanisms is possible only by analyzing the evolution patterns of the fine structure parameters and assessing the contributions of structural components and defective substructure to rail hardening during long-term operation. At present, this is possible with the use of highly informative methods of transmission electron microscopy (TEM), which allow us to carry out a comprehensive analysis of both morphology and defect structure, as well as phase composition with a sufficient degree of the locality over the rail cross-section.

The main areas of research on worn-out rails in the world scientific community can be divided into three large interrelated and complementary groups:

1. Formation and behavior of structural-phase states and nanoscale structures, wear of rail steel, and related processes during long-term operation.

2. Modelling of the processes occurring in the surface layers of rails under severe plastic deformation (megaplastic deformation (MPD)).

3. Methods and techniques for assessing the structural and phase state of rails, internal stresses, and their evolution during the life cycle.

The behavior of rail metal during long-term operation in many respects corresponds to the processes of megaplastic deformation, which ensure the production of bulk nanostructured metallic materials.

At modern train speeds and high contact pressures (≥ 1 GPa), there is a strong refinement of the microstructure to the nanoscale range and even the formation of structureless

Structure and Properties of Lengthy Rails after Extreme Long-Term Operation
Materials Research Foundations **106** (2021)

Materials Research Forum LLC
https://doi.org/10.21741/9781644901472

"white" layers. These processes take place in the surface layers of the rails even with a relatively small hauled load of 100-500 million gross tons.

It is well known by now that a high level of properties of nanomaterials is associated with the presence of a large number of interfaces, residual stresses, defects, boundary segregations, and non-equilibrium phases. However, the same factors lead to an excess of free energy. It is quite obvious that with prolonged deformation effects, recrystallization, segregation, homogenization, and relaxation processes can be initiated. Besides, phase transitions, decomposition and formation of phases, amorphization, sintering, and filling of micro- and nanopores (nanocapillaries) might also take place. These processes significantly affect the evolution of the nanostructure, sometimes even leading to its disappearance, and are often accompanied by a deterioration in physical, mechanical, and chemical properties [1-3].

In this regard, a joint analysis of the behavior of the structure, phase composition, and defective substructure at different distances from the rolling surface along the central axis and the fillet during long-term operation seems to be very desirable.

The authors consider it their duty to express their sincere gratitude to the staff of Siberian State Industrial University, Institute of High Current Electronics, Siberian Branch of the Russian Academy of Sciences, National Research Tomsk Polytechnic University, Tomsk State University of Architecture and Civil Engineering, JSC Evraz-ZSMK for their help in conducting experiments, to Doctor of Physics and Mathematics, Professor V.I. Danilov, and Doctor of Technical Sciences, Professor V.V. Muravyov for reviewing the book and making comments.

1. Evolution of Structure and Properties of Rails During Operation

1.1 Structural-phase state and properties of rails during long-term operation

1.1.1 Influence of various factors

Rail life is determined by many factors: metal purity, internal structure, phase composition, operating conditions, heat treatment technology, etc. At modern train speeds and high contact pressure, even with a relatively low transported tonnage, a strong change is observed in the structure of the surface layers of the rails, an abnormally high value of microhardness and the phenomenon of cementite decomposition are noted.

The phenomenon of deformation-induced decomposition of cementite, which is very stable under normal conditions, forces us to closely consider nanoscale systems, where phase equilibrium can be violated. By comparison with grains of ordinary size under severe plastic deformation, the inverse transformation of the bcc lattice of ferrite into the fcc lattice of austenite was observed [4]. Nanocrystalline grains of austenite, 10-20 nm in diameter, are formed in a ferrite matrix according to the Kurdyumov-Sachs orientation relationship. These results are in good agreement with the predictions of molecular dynamics models. The increase in the Gibbs free energy of the ferrite due to the formation of the nanocrystalline structure and the dissolution of carbides along with the high level of shear stress provide the driving force for this reverse martensitic transformation.

Deformation-induced austenite is an unstable phase at room temperature and can usually turn into ferrite upon unloading, however, in a different crystallographic orientation. This stress-induced phase transformation provides a new mechanism for grain slippage in the nanostructured state. However, there are other factors to consider. As noted in [1-3], during torsion under high pressure at 350 °C of medium carbon steel (0.45% C) and Armco iron, the formation of an ultrafine-grained structure with high-angle boundaries is noted. Compared to Armco iron, medium-carbon steel gains high values of strength and plasticity. According to the authors of [5], this is due to the refinement of the grain size and the formation of highly fragmented cementite particles.

The formation of a nanostructure, which is far from equilibrium, in high-carbon steel (0.6-0.8% C) was considered in [6] under MPD. The resulting nanostructured material has increased thermal stability contrary to the conventional martensite obtained by quenching from elevated temperatures. The microhardness of the nanostructured surface, 9.4 GPa, slightly exceeds the value for martensite. It practically does not change up to

350 °C. The decrease in microhardness with a temperature above 350 °C is associated with the precipitation of the ε-carbide phase. After annealing for one hour at 600 °C, a deceleration of the growth of ferrite grains at the level of 700 nm was observed [6].

It is shown in monographs [1-3] that torsion of rail steel under a pressure of 7 GPa leads to the formation of a highly deformed misoriented pearlite structure. Cementite plates are divided into nanofragments 5–20 nm in length; in several areas, carbon transfer is observed along dislocations of the cellular-network structure [7].

Phase transformations in pearlitic steel due to a high level of stresses and deformations are considered in [1-3, 8]. A high level of pressure and shear deformations leads to the observed transformations, which exclude the temperature effect. One of them is the deformation-induced decomposition of cementite, which is associated with the presence of friction forces at the solid cementite-ferrite interface. At high stress and strain levels, the ferrite matrix behaves like a viscoelastic fluid. Friction at the particle-matrix interface provides two effects. The first one is the formation of a high level of deformations on the precipitates, which in turn leads to a shift in thermodynamic equilibrium and dissolution of cementite. The second effect is the abrasion of the cementite phase due to friction at the interphase boundary and the mechanically induced drag of carbon atoms by the ferrite [1-3]. Recent studies have confirmed that this may lead to the formation of non-stoichiometric cementite [8]. The existing theories of the transfer of atoms by moving dislocations can be considered only as one of many possible mechanisms. In this respect, the process can be called athermal, where the temperature indirectly affects the transfer processes. As in [4], a stress-induced α → γ transformation was noted, which was never observed during the deformation of coarse-grained steels with insignificant degrees.

1.1.2 Formation of nanostructures

The mechanisms of the formation of nanostructured-phase states during severe plastic deformation by torsion under high pressure have been analyzed in some works [1-3, 9] for pearlitic steel (0.86% C). Ferrite grinding to a grain size of 10 nm is accompanied by complete dissolution of cementite and an increase in hardness up to 11 GPa. The deformation-induced dissolution of cementite corresponds to a three-stage process of the formation of nanostructures. At shear deformation $\gamma < 100$, about 50% of the carbide phase dissolves. This is accompanied by the formation of a cellular structure of ferrite, elongation of cementite colonies in the shear direction, and thinning of cementite plates. At the second stage, corresponding to $100 < \gamma < 200$, the cementite dissolution rate slows down. At this stage, a nanostructure is formed in the cementite colonies. This stage of cementite dissolution is similar to the formation of a "white" layer on the surface of railway rails [1-3]. Further dissolution of carbides is associated with the expansion of nanostructured regions. At the third stage, $200 < \gamma < 300$, the material is fully nanostructured with a grain size of less than 10 nm. The formation of such a nanostructure with dissolved carbon is not associated with an increase in the lattice parameter, which may indicate that dissolved carbon is deposited at grain boundaries and in the dislocation cores. As in [9], the cementite dissolution mechanism is based on the model of the plastic phase (ferrite) flow under the action of high external stresses and internal ones around solid carbide precipitates. This work also discusses the role of friction forces at the precipitate-matrix interface. Nanostructured pearlitic steel, obtained by the methods of severe plastic deformation, has the mechanical properties of ceramic materials [1-3].

The mechanical properties of nanocrystalline materials are described in detail in reviews [1-3, 10-11]. Their study in nanocrystalline surface layers is fraught with understandable difficulties. Noteworthy is the work [12], in which a hardness value of 14 GPa, according to the Meyer scale, was obtained during the indentation of the rail surface after long-term operation. It made it possible to recalculate the obtained values of hardness according to the formulas proposed in [12] and to estimate the mechanical properties under compression. As the temperature rises, the hardness decreases to 1 GPa at 630 °C. The authors have shown that, at temperatures below 150 °C, the nanocrystalline surface layer cannot deform plastically. If the temperature exceeds 500 °C at the surface, then a dynamic return is possible. Hence, the authors do not associate the reason for the growth of the nanocrystalline layer with temperature effects [1-3].

The most comprehensive analysis of the mechanical properties of nanomaterials is presented in the review [13]. However, despite such extensive information, a comprehensive consideration of the problem of the mechanical behavior of nanostructures, containing the analysis of data not only on strength and plasticity, but also on fracture toughness, creep, superplasticity, and other mechanical characteristics for metallic nanomaterials, including materials containing brittle phases, is far from completion and seems to be very relevant [1-3].

The authors of [2] also noted that a detailed systematization of defect structures that arise in various materials with an increase in the degree of plastic deformation was carried out by E.V. Kozlov and N.A. Koneva [14]. They showed that, when approaching the region of megaplastic deformation (MPD), depending on the nature of the material, a sequential change from one structural state to another (cellular, strip, fragmented structures, etc.) occurs, similar to structural phase transitions. In this case, internal stresses and conditions for the manifestation of anomalies in the mechanical behavior of crystals change. Here, the works [1–3, 15] should be mention, which show that with an increase in the degree of

MPD, a very large number of excess point defects (mainly vacancies) are formed in the structure, which can stimulate the diffusion phase transformations during deformation.

The specificity of the behavior of nanostructures under extreme conditions lies, first of all, in the fact that these objects (except for supramolecular systems) in the overwhelming majority of cases are essentially non-equilibrium [2, 16]. Numerous interfaces in the form of intergrain and interphase boundaries, triple and quadruple joints, as well as residual stresses and defects, segregations, and non-equilibrium phases - all this, on the one hand, stimulates the appearance of a high level of physical and mechanical properties of nanomaterials (in comparison with conventional large-crystalline objects). However, on the other hand, the presence of an excess of free energy in nanostructures also imposes increased requirements on their stability [1-3]. It is quite obvious that under thermal, deformation and corrosive effects, as well as under radiation conditions, recrystallization, segregation, homogenization and relaxation processes, phase transitions, decomposition and formation of phases, amorphization, sintering, and flooding of micro- and nanopores (nanocapillaries) can take place. To one degree or another, all this can lead to the evolution of the nanostructure, and sometimes to its annihilation, accompanied by a change (deterioration) in its physical, mechanical, chemical, and biological properties. Therefore, the identification of the regularities of the stability of nanostructures acquires special relevance, being one of the main problems of modern nanostructured materials science [2, 16].

In the process of severe plastic deformation, which occurs during the long-term operation of rails, the surface layers are transformed into a nanocrystalline α-Fe-C alloy [17] with an inhomogeneous microstructure and a grain size of 15–150 nm. The nanostructure of the rail surface is the result of processes in the solid phase similar to mechanical alloying. Dissolution of cementite occurs at a temperature below the austenitizing temperature. The driving force for the dissolution of carbides is a high level of stresses acting on precipitates and non-equilibrium interphase states of ferrite-cementite [1-3].

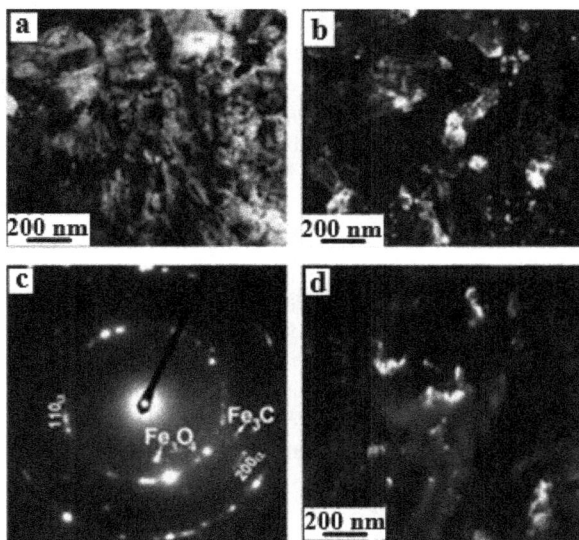

Figure 1.1 *TEM image of the microstructure of the "white layer" formed after an operating load of 3.6 × 10⁸ tons (sample I): (a) bright-field image;(b) dark-field image in the reflection (110) α - Fe; (c) micro-electronogram with the indicated and indexed ferrite rings, the reflections of cementite and iron oxide are indicated; (d) dark-field image created using cementite reflections.*

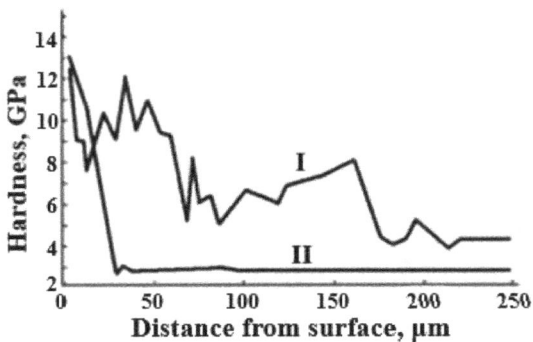

Figure 1.2 *Hardness profiles for rails after 3.6 × 10⁸ tons (sample I) and 3.8 × 10⁶ tons (sample II) operating load [18].*

At modern speeds of trains and high contact stresses, even with a relatively small hauled tonnage, the formation of structureless regions to a depth of up to 100 μm is observed in the surface layers of rails and wheels. The analysis of structural-phase changes in rails under severe plastic deformation corresponding to 3.6×10^8 and 3.8×10^6 gross tons (Fig. 1.1) is carried out in [18]. In both cases, an abnormally high microhardness value of up to 12 GPa is found on the surface (Fig. 1.2). This corresponds to the observed microhardness value for the "white" layer. In this layer, the microstructure is extremely heterogeneous, grain sizes have a bimodal distribution and vary in the range from 20 to 500 nm [1-3]. With such intense deformations, cementite plates are either bent or destroyed; an extremely high density of dislocations is found at the ferrite-cementite interface. The microstructure of the "white" layer zone is in many respects similar to the structure observed under conditions of severe plastic deformation during equal-channel angular pressing and shear torsion. Partial dissolution of cementite and the formation of austenite were shown in [18]. According to the authors, the dissolution of cementite occurs similar to mechanical alloying of steel, and austenite appears due to the reverse γ → α transformation.

1.1.3 White layer on the rolling surface of the rails

It has been shown in [1-3] that the white layer (WEL) [19-26], which is a result of rail-to-wheel contact, is a common microstructural feature that occurs on rail surfaces and is often associated with rolling contact fatigue damage (RCF). During the movement of trains, a part of the original pearlite microstructure on the rail surface transforms into another structure. The new structure usually appears as a 10–20 μm thick layer, which, after etching with a nitric acid solution in ethanol, turns white under an optical microscope. Besides, a white layer is a common problem in rails with a pearlite microstructure since this layer occurs: (1) with different initial microstructural components, i.e. ferrite-pearlite [27] or completely lamellar pearlite [1-3, 19, 23, 24], and (2) on various sections of the track, for example, tangential rails or curved rails. The presence of a white layer can lead to the appearance of cracks after brittle fracture of the layer, which is due to its high hardness (up to 1200 HV) [23]. Worldwide observations of crack initiation on a white layer have been presented in metallographic studies of loaded rails in various countries, including Japan [27], the Netherlands [28], Australia [29], and the United Kingdom [30]. It is noted in [1-3] that Carrol et al. [31, 32] determined the effect of the white layer on the behavior of RCF in rails using laboratory tests of double discs and suggested that the white layer facilitates both wear and RCF processes in rails. In their subsequent numerical analysis, the concentration of stresses and strains in the white layer was estimated, which coincided with their experimental observations. The origin of the microstructure containing the white layer is usually determined using a combination of advanced research methods [1-3]. The detection of tetragonal deviation from the body-centered cubic lattice (BCC) and retained austenite using X-ray diffraction is usually taken as convincing evidence of the presence of martensite in the white layer [21, 33]. Additional evidence includes a martensitic twinning substructure observed by transmission electron microscopy (TEM) [21]. Using TEM, Takahashi et al. [24] determined that the white layer has a high dislocation density and consists of grains of several hundred nanometers in size, which are much thinner than pearlite under the white layer. The TEM study was combined with an atomic probe tomography (APT) analysis to quantify the diffusion of manganese in the white layer, and the authors concluded that the rail temperature increased to at least 900 °C. Accordingly, a hypothesis was proposed that the formation of a white layer occurs through a martensitic phase transformation [1-3].

Another hypothesis is that the white layer consists of nanocrystalline ferrite grains, due to deformation accumulated from the passing train wheels. The initial pearlite-ferrite structure undergoes significant refinement to grain sizes of tens of nanometers in dimension, and cementite dissolves in this structure due to plastic deformation. The extremely high hardness [25, 34] is considered abnormal for the martensitic microstructure obtained by standard heat treatment, but it can be attributed to hardening due to ultrafine ferrite grains. In some cases, the formation of a white layer is attributed to this hypothesis, despite the appearance of martensite-like characteristics, for example, in a study based on TEM. Baumann et al. [20] reported a completely martensitic microstructure of the white layer without retained austenite. The corresponding calculated temperature significantly differed from the temperature required for the transformation of pearlite into austenite [1-3]. Consequently, they attributed the formation of the white layer to repeated deformation. The formation of the white layer, following the above hypotheses, is firmly supported by individual laboratory models. For example, (ultra) fast processing can lead to martensite with hardness and grain sizes comparable to those of the investigated white layer in the rails [35], or martensite with high hardness and double substructure [21]. A symmetrical white layer is also often observed on the surface of loaded parts subject to significant temperature increases, for example, during surface treatment [1-3, 33, 36-38]. However, in pearlite subjected to severe plastic deformation, nanocrystalline ferrite white layers are formed [25, 38].

The extremely high hardness of the white layer and its constituent nanometer grains [25] can be reproduced using intensive plastic deformation. Some critical deformation may be formed to produce nanoscale grains and complete dissolution of cementite to be achieved. Consequently, the microstructure of the white layer must be correlated with the loading conditions of the investigated rail specimen. Concerning the contact of the wheel with the rail, the following assumptions are considered valid: (1) the white layer is formed as a result of combined plastic deformation and temperature increase, and (2) individual contributions depend on the loading history of the rails [1-3]. However, automatic orientation mapping techniques such as electron backscattered diffraction (EBSD) and the recently developed automatic transmission electron microscopy (ACOM-TEM) orientation mapping with improved spatial resolution [39, 40] have been used in only a few of the above studies.

The EBSD and ACOM-TEM methods provide accurate measurements of the orientation and misorientation of crystals and are inherently useful for detecting deformation structures in the white layer [35], and the deformation gradient in rails [41, 42]. The characterization and quantification of such structures using other methods, such as TEM and APT, is very difficult. Besides, EBSD can scan a large volumetric sample, including the white layer and the surrounding matrix [1-3]. Therefore, compared to TEM and APT, EBSD provides a broader view of the differences between (e.g., grain size, deformation scale) WEL and the pearlite below it. The spatial resolution of EBSD, which is typically 50–100 nm, can limit the identification of very small grains in the white layer [42].

A very small grain size (50 nm) was detected by the Kikuchi transmission diffraction (TKD) method in a white layer formed after substantial plastic deformation in high-carbon steel (one mass percentage of carbon C) with the initial microstructure of martensite and 30-40% residual austenite [43]. Alternatively, additional information can be obtained using the ACOM-TEM method, where a spatial resolution of 2 nm has been reported for field TEM radiation [39]. Therefore, EBSD and ACOM-TEM will be useful for characterizing the white layer and additional information can be obtained from crystallographic features. In this paper, the authors characterize the white layer observed in the R260 Mn rail using different methods associated with different levels of resolution. It is noted that the mechanism of formation of a white layer is investigated by evaluating crystallographic, morphological, and compositional characteristics [1-3]. Methods such as XRD, EBSD, and ACOM-TEM are used to identify the phase components of the white layer. Also, EBSD and ACOM-TEM are used to display the phase distribution and detect local misorientation changes caused by the passage of the train. TEM will be used to characterize morphological and phase components of the white layer at a higher resolution level than XRD and EBSD. Compositional variations in the white layer are revealed through APT. Finally, based on the results obtained, the mechanism of the formation of the white layer is discussed.

Figure 1.3 *White layer on the rail surface after operation [44].*

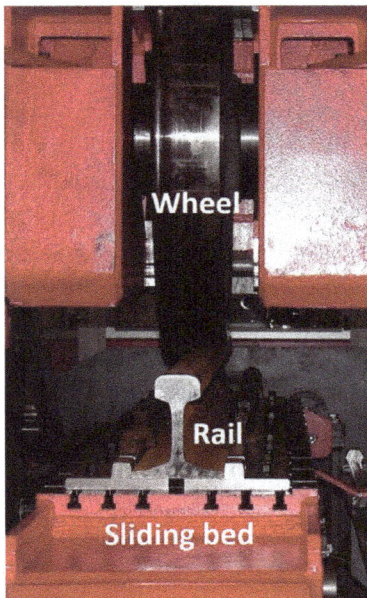

Figure 1.4 *Schematic of the experimental setup. Speeds up to 2 m/s, normal load up to 40 tons [44].*

The formation of a martensitic white layer is possible in two cases: 1) at intensive and plastic deformation (MPD) at temperatures below the austenitizing temperature (friction martensite); 2) when heated above the austenitizing temperature and subsequent rapid cooling (thermal martensite) [1-3] (Fig. 1.3). The full-scale experiments performed on the equipment of Voestalpine Schiene GmbH (Austria) (Fig. 1.4) confirmed the predictions of the model [44] and the validity of the second scenario of the formation of a martensite white layer. The importance of these results lies in the fact that the formation of ''squat'' type contact fatigue defects during the operation is preceded by the formation of a white layer [44], and the model predicts its appearance depending on the velocity, axle load, and contact area.

1.1.4   Structure and properties of body-hardened worn rails

In works [1-3, 45-60], the methods of modern physical materials science were used to study the structure, phase composition of defective substructure, microhardness, and tribological properties, which are formed at various distances along the central axis and along the fillet in the head of body-hardened rails after different periods of operation (500 and 1000 million tons of hauled tonnage).

Let us dwell in more detail on the results of these works, since they are directly related to the considered problem of differentially hardened rails.

In the initial state of the rails, the following structural components were identified: colonies of lamellar pearlite (relative content ≈ 0.7), grains of a ferrite-carbide mixture (≈ 0.25), and grains of structurally free ferrite (≈ 0.05).

Central axis changes (500 million tonnes)

The results of tribological tests of the metal of the rolling surface of rail steel, presented in Table 1.1, show that the rail operation leads to a significant decrease in wear resistance. After the hauled load of 500 million gross tons, the wear resistance of rails decreases by approximately three times. The decrease in wear resistance is accompanied by a decrease in the coefficient of friction. The hardness of the surface layer of the metal of the rails after the hauled load of 500 million gross tons is approximately 7.0 GPa. With increasing distance from the rolling surface, the hardness of the metal decreases, reaching a plateau at a distance of 10 to12 mm.

Table 1.1 *Results of tribological tests of the metal of the rolling surface of rails.*

Designation	Friction coefficient, μ	Wear rate, 10^{-5}, $mm^3 / N \cdot m$
Initial state	0.49	3.2
500 million tons	0.36	9.8
1000 million tons	0.43	10.9

Three characteristic zones are revealed on the fracture surface: a zone of normal crack growth, a fracture zone, and a zone of accelerated crack growth separating them. A mixed destruction mechanism is implemented. Viscous fracture pits and quasi-cleavage facets are found. Pits are the predominant element of the fracture surface structure and are formed as a result of cutting off micropores through which the destruction of grains has passed.

The exploitation of steel is accompanied by a significant change in the structure of the surface layer. After the hauled tonnage of 500 million tons, the destruction of cementite plates of pearlite colonies is observed (Fig. 1.5). Rounded cementite particles are detected amidst lamellar pearlite colonies, the sizes of which are 30 - 50 nm (Fig. 1.5, a) and 10 - 15 nm (Fig. 1.5, b). Fracture of cementite plates is accompanied by fragmentation of the ferritic component of pearlite (Fig. 1.5, a). The average fragment size is 150 nm. A dislocation substructure is observed among the fragments; the scalar dislocation density reaches 1×10^{11} cm^{-2} [2, 45, 50-54].

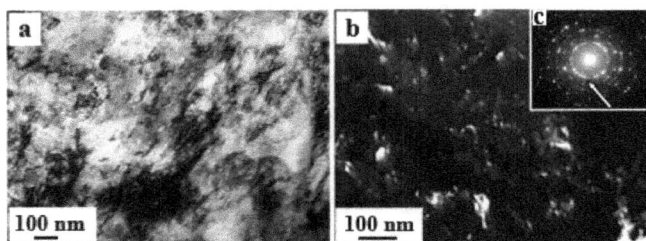

Figure 1.5 *Electron microscopic image of the structure of the surface layer of rail steel. (a) bright field; (b) dark-field obtained in the reflection [112] Fe$_3$C; (c) micro-electronogram; the arrow indicates the reflex in which the dark field was obtained [1, 2].*

The facts obtained may indicate the occurrence of two competing processes during the exploitation of steel: (1) the process of cutting cementite particles with their subsequent movement into the volume of ferrite grains or plates (in the structure of pearlite); (2) the process of cutting, the subsequent dissolution of cementite particles, the transfer of carbon atoms to dislocations (in the Cottrell atmosphere), the transfer of carbon atoms by dislocations into the bulk of ferrite grains (or plates) with the subsequent re-formation of nanosized cementite particles.

At a depth of 2 mm from the rolling surface, cementite plates are broken into a set of separately located particles of globular morphology. Particle sizes vary from 5 to 13 nm.

At the same time, cementite particles, the size of which is 3 - 5 nm, are also detected in ferrite plates. This indicates the dynamic aging of steel during the operation of the rails.

Fillet changes (500 million tonnes)

As noted in [1-3], in contrast to the increase in hardness along the central axis, the operation of rails leads to some softening of the surface layer. The most significant transformations of the grain structure of rail steel during long-term operation are observed in the surface layer. The evolution of the structural-phase state of pearlite lamellar morphology is the dissolution of cementite plates. This leads to the formation of a chain of globular carbide phase particles in place of the cementite plate, which is possible due to the escape of carbon atoms from the cementite crystal lattice to dislocations. The second stage of this transformation is the formation of nanosized particles of the carbide phase in the ferrite interlayers of the pearlite colony. The evolution of the structural-phase state of the grains of the ferrite-carbide mixture is accompanied by the formation of a fragmented substructure with the size of fragments (subgrains) 250 - 300 nm. Particles of the second phase are located in the volume and along the boundaries of the fragments. Judging by the micro-electron diffraction patterns, the particles of the second phase are iron carbides; in some cases, reflexes of iron oxides are detected.

At a depth of 2 mm, the operation of the rails led to a significant transformation of the substructure. First, the scalar density of dislocations located in the volume of the ferrite component of the material increased by 1.5 - 2 times. Secondly, fragmentation and destruction of cementite of lamellar morphology is recorded. Third, the formation of nanosized particles of the carbide phase in the ferrite component of steel is observed. Nanosized particles are detected in pearlite grains, in grains of a ferrite-carbide mixture, and grains of structurally free ferrite.

This fact indicates a multistage process in steel during the operation: the dissolution of cementite particles of the initial state, transition of carbon atoms to dislocations (in the Cottrell atmosphere and dislocation cores), transfer of carbon atoms by dislocations into the volume of ferrite grains or ferrite interlayers, recurring release of carbon atoms with the formation of nano-sized particles of cementite of rounded shape.

It has been established that, regardless of the distance to the rolling surface, the main source of curvature-torsion of the crystal lattice of the metal is the interphase boundaries, i.e. the interfaces of cementite and ferrite (Fig. 1.6). It is observed an increase in the number of bending extinction contours per unit surface area of the material (the number of stress concentrators increases) and a decrease in the transverse dimensions of the contours (the amplitude of long-range internal stress fields increases) when nearing the rail fillet surface [2, 45, 50-54].

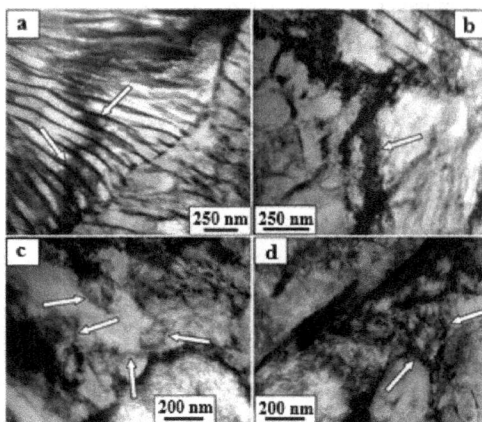

Figure 1.6 *Electron microscopic image of the fillet rail structure after the hauled load of 500 million gross tons. (a, b): layer at a distance of 2 mm from the surface of the fillet; (c, d): the surface layer of the fillet. Arrows indicate bending extinction contours [1, 2].*

Central axis changes (1000 million tons)

As well as after the hauled load of 500 million tons, a decrease in the wear resistance of the rolling surface was noted with a slight decrease in the coefficient of friction (Table 1.1). A twofold decrease in hardness is noted in the near-surface layer with a thickness of approximately 2 mm to the steel layer located at a distance of about 10 mm. The latter indicates the degradation of the material structure during the operation [1-3].

As well as after the hauled load of 500 million gross tons, three characteristic zones were identified: the zone of normal crack growth, the break zone, and the zone of accelerated crack growth separating them. The thickness of the fracture zone varies within 200 - 300 microns, the zone of accelerated crack growth is 70 - 90 microns. As the distance from the zone of an accelerated crack is growing, the cellular nature of the fracture is accompanied by the appearance of grooves, the fraction of the surface area on which the fracture occurred by the cleavage mechanism increases, and smooth cleavage becomes predominant. Judging by the size of the cleavage areas, the destruction of the material is intergranular at this stage of crack growth.

Comparison of the thickness of the break zone for different periods of operation shows that it practically does not depend on the hauled tonnage. A detailed analysis of the break zone made it possible to identify a certain surface sublayer with a thickness of up to 40 microns. The surface sublayer is characterized by the presence of a large number of microcracks, micropores, and potholes. The sizes of micropores vary within 1 - 2 microns. The formation of a fracture can lead to the peeling of the surface layer from the bulk of the sample.

If, after hauled 500 million gross tons, a strip substructure is formed in the volume of ferrite grains, then after 1000 million gross tons, a predominantly subgrain structure is revealed in ferrite grains, which indicates the initial stage of dynamic recrystallization of the material. Chaotically located dislocations are found in the volume of subgrains, the scalar density of which does not exceed 10^8 cm^{-2}. The next difference is the formation of a structure in the surface layer, the micro-electron diffraction patterns of which have an extraordinary appearance. Namely, they contain separately located point reflections belonging to the α-phase (solid solution based on bcc iron), and a large number of thin diffraction rings, apparently belonging to nanosized particles of carbide and oxycarbide phases. It can be assumed that the destruction of lamellar pearlite colonies and dynamic recrystallization of ferrite grains can contribute to a decrease in the hardness of the surface layer of steel up to the value of the hardness of the initial state [2, 45, 51-54].

The structure of pearlite changes from a state close to the initial one to a state when cementite plates are detected only by the contrast of the image and the increased density of nanosized particles of the carbide phase. That is, the transformation consists in the dissolution of cementite plates due to the escape of carbon atoms from the crystal lattice of cementite at the dislocation (Cottrell atmospheres are formed, the area of the dislocation core is enriched with carbon atoms).

The transformation of grains of a ferrite-carbide mixture also consists in the formation of a fragmented (subgrain) structure. An interesting feature of this transformation is the redistribution of cementite particles. If the grain structure of the ferrite-carbide mixture of the initial steel was characterized by an almost uniform distribution of globular cementite particles over the grain volume, then in the layer located at a distance of about 2 mm from the rolling surface the particles are located mainly along the subgrain boundaries. We can assume two options for the formation of such a structure. Firstly, the formation of sub-boundaries is tied to particles of the carbide phase and, secondly, the dissolution of particles located in the volume and their re-precipitation at the boundaries of subgrains is possible.

Fillet changes (1000 million tonnes)

A significant (1.5-2 times) increase in the microhardness of a layer with a thickness of up to 10 mm and the evident formation of a hardened surface layer is noted as compared to the results obtained after the hauled load of 500 million tons.

The main type of surface layer structure is a fragmented substructure with fragment (subgrain) sizes of 100 to 150 nm. The nanoscale state of the structure of the surface layer of steel is also confirmed by the quasi-ring structure of the micro-electron diffraction pattern. The boundaries of the fragments are decorated with particles of the second phase. The particles have a round shape, and their sizes vary within 15 - 20 nm. The indexing of electron diffraction patterns shows that the particles of the second phase are iron carbides, and in some cases, reflexes of iron oxides are detected.

At a depth of 2 mm from the surface of the fillet of the rails, after the hauled load of 1000 million tons, the formation of a substantially inhomogeneous structure is observed. First, grains of lamellar pearlite are determined, in which ferrite plates are broken into misoriented regions. A similar structure is formed in the grains of the ferrite-carbide mixture. Secondly, grains of pearlite and grains of a ferrite-carbide mixture with partial or complete dissolution of cementite plates were found. In this case, in place of the cementite plates, a certain set of round-shaped carbide phase particles with sizes 15 - 30 nm is formed. Third, there are grains of structurally free ferrite containing a subgrain structure. At a distance of about 10 mm from the fillet surface, the structure of rail steel in morphological and phase composition is close to the structure of rail steel before the operation.

Having determined the quantitative characteristics of the steel structure, in the first approximation, an analysis of the physical mechanisms responsible for the evolution of steel hardness during the operation of rails is carried out. The contributions due to the friction of the matrix lattice, intraphase boundaries, dislocation substructure, long-range stress fields, and particles of the carbide phase were taken into account [2, 45].

Fillet surface

From the results presented in Table 1.2, it follows that the parameters of the defective structure depend on the operating time of the rails. Using these results of quantitative analysis and the known equations of contributions to hardening from the available literature, the mechanisms of hardening the surface of the working fillet were evaluated (Table 1.3). It can be seen that in the volume of the material adjacent to the working surface, the main mechanisms of material hardening are the substructural mechanism (steel fragmentation, the formation of subboundaries) and the mechanism caused by elastic distortions of the crystal lattice of the material due to the incompatibility of deformation of neighboring grains, pearlite colonies, subgrains, inclusions of the carbide phase, and α-matrices. Note that the elastic component of stress fields is approximately five times higher than the yield stress of rail steel, equal to about 900 MPa. This should lead to plastic relaxation of internal stresses. Taking into account the given fact, let us limit the value of internal stress fields by the value of the yield strength of steel. The results of such adjustment of theoretical estimates of the rail metal yield strength are given in parentheses in Table 1.3 [2, 45].

Table 1.2 Characteristics of a defective substructure formed during the operation of rail steel in a layer located at the surface of the "working" fillet.

Rail operating mode	500 million tons	1000 million tons
L, nm	520	60
σ_{az}, degrees	4	7
$<\rho> \times 10^{-10}$, cm^{-2}	4.2	5.1
$\rho_\pm \times 10^{-10}$, cm^{-2}	3.3	5.1
χ_{pl}, cm^{-1}	944	1276
χ_{el}, cm^{-1}	154	5707

Notations: L is the average size of fragments, σ_{az} is the azimuthal component of the total angle of disorientation of fragments, $< \rho >$ is the scalar density of dislocations, $\rho_\pm$ is the excess density of dislocations, χ_{pl} is the plastic component of the curvature-torsion gradient of the crystal lattice; χ_{el} is the elastic component of the curvature-torsion gradient of the crystal lattice.

Table 1.3 Estimates of the hardening mechanisms of the rail steel layer located at the surface of the "working" fillet.

Hardening mechanism / Rail operating mode	500 million tons	1000 million tons
Substructural hardening, $\sigma(L)$, MPa	288	2500
Dislocation hardening, σ_∂, MPa	205	226
Strengthening by fields of internal stresses:		
Plastic component, σ_{pl}, MPa	182	226
Elastic component, σ_{el}, MPa	123	4565 (900)
Strengthening the pearlite component, $\sigma(P)$, MPa	82	0
Strengthening with cementite particles, σ_{cp}, MPa	0	260
Additive summation, MPa	880	7777 (4100)

Rolling surface (center axis)

The results of the quantitative analysis of the defective substructure of the rolling surface and the assessment of the contributions of structural elements to the yield stress are

Structure and Properties of Lengthy Rails after Extreme Long-Term Operation
Materials Research Foundations **106** (2021)

Materials Research Forum LLC
https://doi.org/10.21741/9781644901472

presented in Tables 1.4 and 1.5. It is seen that the main hardening mechanism is substructural, due to the formation of subboundaries.

Table 1.4 *Characteristics of a defective substructure formed during the operation of rail steel in a layer located at the rolling surface.*

Rail operating mode	500 million tons	1000 million tons
L, nm	150	330
σ_{az}, degrees	7	4.6
$<\rho> \times 10^{-10}$, cm^{-2}	10.0	5.9
$\rho_{\pm} \times 10^{-10}$, cm^{-2}	9.8	5.6
χ_{pl}, cm^{-1}	2490	1408
χ_{el}, cm^{-1}	0	127

Table 1.5 *Estimates of the hardening mechanisms of the rail steel layer located at the rolling surface.*

Hardening mechanism Rail operating mode	500 million tons	1000 million tons
Substructural hardening, $\sigma(L)$, MPa	1000	455
Dislocation hardening, σ_{∂}, MPa	316	243
Strengthening by fields of internal stresses:		
Plastic component, σ_{pl}, MPa	316	236
Elastic component, σ_{el}, MPa	0	102
Strengthening of perlite component, $\sigma(P)$, MPa	0	0
Strengthening with cementite particles, σ_{cp}, MPa	260	260
Additive summation, MPa	1892	1296

Analyzing the results given in Tables 1.4 and 1.5, and Fig. 1.7 (a), we note that the additive yield stress of rails in the rolling surface area after the hauled load of 500 and 1000 million gross tons changes in a similar way, namely, reaches maximum values on the rolling surface and decreases rapidly with increasing distance from the rolling surface along the centerline. At the same time, the additive yield point of the rail metal after the hauled load of 500 million gross tons is higher than the yield point of the rail metal, which is formed after the hauled load of 1000 million gross tons.

The microhardness profile of the metal after the hauled load of 500 million gross tons changes similarly to the change in the theoretically calculated yield stress of the

Materials Research Forum LLC
https://doi.org/10.21741/9781644901472

corresponding material (Fig. 1.7, (b), curve 1). The change in the value of microhardness, measured in the experiment, at the extreme points of the profile is approximately 1.5; in the case of the yield point, determined based on theoretical estimates, it is about 2.2.

An analysis of the microhardness profiles demonstrated an interesting feature. An increase in the hauled load up to 1000 million gross tons leads to some (by a factor of about 1.4 times) weakening of the surface layer of the metal of the rolling surface of the rails (see Fig. 1.7, (b), curve 2).

Based on the analysis of the impact fracture surface of rail samples, it was shown in [2, 50, 53] that operation (1000 million tons) leads to the formation of a surface layer characterized by a large number of micropores and microcracks. These facts, on the one hand, can be one of the main reasons for the softening of the surface layer of the rolling surface of the rails, observed during the construction of the microhardness profile, and, on the other hand, can be an explanation of the reported disagreement between the experimentally measured and theoretically calculated results (see Fig. 1.7), which are based on studies of the defective substructure and phase composition of the steel.

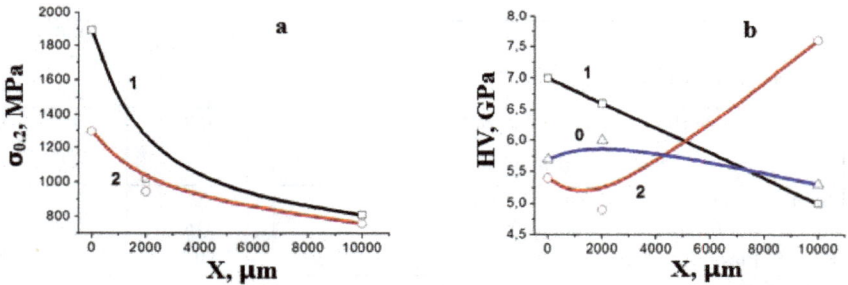

Figure 1.7 *Dependence on the distance X from the rolling surface of the rail of the average microhardness HV (b) and the theoretically calculated yield point (a). Curve 0 is the initial state of the rails, 1 - after the hauled tonnage of 500 million tons, 2 - 1000 million tons [1, 2].*

The volume gradient of microstructures in worn-out R260 rails has been studied by optical and scanning microscopy, micro-indentation, and backscattered electron diffraction [61]. It was shown that perlite colonies are fragmented to a depth of 3 mm. Thinning of cementite plates in shear directions was observed on the entire rolling surface. In the transitional layer (3-4 mm from the surface), not fragmented colonies of perlite are found. In fragmented colonies, cementite plates are bent and destroyed, and, accordingly, the growth of high-angle boundaries is observed. As the rolling surface is approached, the distance between the plates decreases. The research results are of practical value because of the contribution of the gradient structure to the ductility of rails and the dissolution of fatigue cracks.

Among the few publications devoted to structural transformations in the surface layers of rails, the work of Czech researchers [62] stands out, in which surface defects, layer thickness, hardness, stress state, and microstructure of rails were analyzed after 20 years of operation using highly sensitive methods of non-destructive testing. It is shown that a high density of dislocations and vacancies is formed in the surface layers, forming clusters. Over a long period of operational time, multiple phase transformations occur due to intense cyclic plastic deformations.

The problem of wear in the wheel-rail system is one of the key scientific and industrial concerns. Complex processes in the metal of the rails and the process of long-term operation cannot be predicted. In this regard, safe operating periods pose scientific problems for researchers, on the solution of which the stable operation of entire industries depends on the solution of these problems. Among the latest works in this field, it is worth mentioning publications [63-66], which touched upon a whole range of topical problems.

1.2   Simulation of processes during operation of rails and megaplastic deformation

It was noted in monographs [1-3] that more than one billion dollars are spent annually on Chinese railways due to wear and contact fatigue of rails and wheels. It is well known that wear and fatigue appear on curved track sections and at rail connections. When a wheel moves on a rail and when sliding, the material is removed from the surface of a wheel-rail pair due to high contact and temperature stresses. This is especially the case on heavily loaded road sections [67]. Using a numerical method based on the Kalker contact theory, a wear model is proposed in the "wheel-rail" system. The model takes into account the speed on curved sections, curvature, and slope of the track. This review compares various results with those of numerous other studies. The main findings of the review boil down to the following: 1) The difference between normal loads on the left and right wheels increases with increasing speed on curved sections, which leads to increased wear. The fluctuation of the maximum normal stresses increases with an increase in speed and a change in the profile of the "wheel-rail" system. 2) It is possible to achieve an improvement in motion, contact stresses, and a decrease in rail wear by improving the lifting of the curved section of the track.

A model using finite element methods and the provisions of the theory of plasticity makes it possible to analyze elastoplastic contacts and stresses in the wheel-rail system

Structure and Properties of Lengthy Rails after Extreme Long-Term Operation
Materials Research Foundations **106** (2021)

Materials Research Forum LLC
https://doi.org/10.21741/9781644901472

[68]. The 3D contact was modeled taking into account normal pressure and tangential components. When calculating, the authors used the CONTACT computer package. The results showed that partial slip and creep had a huge impact on residual deformations and, to a lesser extent, residual stresses. At increased speeds and residual deformations, and surface displacements (with an increase in freight traffic), the residual deformations are stabilized after a certain amount of freight traffic. The model's findings also show that changing normal contact forces can lead to plastic deformation induced by a wavy contact surface. The residual stresses and deformations are higher in these areas. Curved track sections represent the greatest difficulty for modeling the processes occurring in the "wheel-rail" system [1-3].

In [69], a numerical model was developed that takes into account the effect of damage in the form of scratches on the rolling surface on the initiation and growth of plastic deformation caused by rail waviness with increasing curvature of the track. The method is based on a combination of the positions of a two-dimensional contact finite element model and a vertical dynamic model of a railway wheelset associated with a curved track section. This takes into account the behavior of the material under repeated contact loads. Waviness tends to move along the direction of travel, and its evolution slows down with increasing traffic. The frequencies of plastic deformation induced by waviness are determined by the frequency parameters of the path itself. Residual stresses stabilize after a certain tonnage. Residual deformations increase to a lesser extent with increasing tonnage [1-3].

A comprehensive approach was developed in [70] to study the probability of fatigue crack initiation with a random distribution of forces in the "wheel-rail" system and random material properties. Initially, the spectrum of loads was established, then the discreteness of fatigue properties was analyzed by the Monte Carlo method. Then three-dimensional finite element models were developed to predict surface stresses at various amplitudes of forces in the wheel-rail system. The proposed probabilistic model takes into account the effects of random forces in the "wheel-rail" system and their influence on the initiation of RCF cracks [1-3].

Vortex structures are formed in the processes of friction and wear occurring during the long-term operation of rails [71-73]. In these works, a mechanism was proposed for the formation of such structures due to the shift of one part of the material relative to another. Its essence lies in the fact that under the action of the friction force, the surface layer of the material moves relative to the underlying layers along the sliding direction. The movement of material, in this case, can be compared with a laminar flow of a viscous fluid, the speed of which is not the same over the cross-section of the flow. Consequently, at different times at different depths, at the boundaries of elastic and plastically deformed

Materials Research Forum LLC
https://doi.org/10.21741/9781644901472

regions, and within the zones of intense plastic shear, there are surfaces of tangential discontinuity of the velocity. From the hydrodynamic point of view, on such surfaces, there is an absolute instability, which is the simplest case of the Kelvin-Helmholtz instability, i.e. the absolute instability of a special type of interfaces separating from each other the flow regions filled with the same or different liquids moving at different speeds [71 -73]. Another surface on which the Kelvin-Helmholtz instability takes place is the boundary between the surface layer and the elastically deformed base material. Analysis of the material flow using the Orr-Sommerfeld equation showed that such instability takes place. This is also confirmed by the data of optical, atomic force microscopy, and scanning electron microscopy (SEM) analysis, which shows the presence of vortex-like structures. In our opinion, this approach will not give a complete picture of the flow of materials at large plastic deformations. The solution to a simpler mathematical problem of the development of the Kelvin-Helmholtz instability must provide a quantitative description of the processes under MPD and long-term operation.

With the help of hydrodynamic representations, it is possible to give a qualitative and quantitative description of the processes of formation of nanostructures under various external influences [1-3]. However, the question of the adequacy of the application of these concepts to solids remains open. Discussion on this issue comes from the classical works [74, 75], where it was suggested that there is a thin quasi-liquid interlayer in the contact of one crystal with another. The presence of this interlayer explains the increased strength of polycrystals upon deformation. On the other hand, as later studies of the structure of grain boundaries showed [76], they consist of misfit dislocations or grain-boundary dislocations; therefore, it is impossible to ignore the structure of grain boundaries when describing the deformation of polycrystals. However, recent studies [77-79] have shown that during megaplastic deformations, the phenomenon of deformational amorphization was discovered, which consists in the transition of materials from a crystalline state to an amorphous state [1-3]. This phenomenon is indicated by the quasi-ring structure of the micro-electron diffraction patterns [77]. The mechanisms of the solid-phase transition "crystal-amorphous state" have not yet been established, but the results [78, 79] allow us to assume that grain boundaries play an important role in the implementation of this transition.

Thus, it should be concluded that the main mechanism for the formation of nanostructures during megaplastic deformation is the flow of materials, which arises as a result of hydrodynamic instabilities.

It has already been noted in [1-3] that in [80-82] a mathematical model of the formation of nanostructures in materials under severe plastic deformation has been developed. It is based on the assumption that under large plastic deformations, the material behaves like a

Materials Research Forum LLC
https://doi.org/10.21741/9781644901472

viscous liquid. All the material was broken into layers that moved at different speeds. According to hydrodynamics, material flow instability occurs on such surfaces. In the presented model, the material was considered as a two-layer liquid with different kinematic viscosities. For each layer, the Navier-Stokes equations and boundary conditions were written. The solution of the resulting system in the form of normal perturbation modes was carried out proceeding from the assumption of the viscous potential flow of the material. As a result, a dispersion equation was obtained, which is the same as the equation obtained earlier for the short wave. Analysis of the dependence of the decrement on the wavenumber shows that it has two maxima, the first maximum falls on the wavenumber corresponding to the micro-wavelength range, and the second to the nanoscale. If we assume that the lower layer is a porous medium, then this is a shift in the values of the critical wavenumbers. The critical wavenumber corresponding to the first maximum increases. The second maximum, on the contrary, is shifted towards the wavenumbers corresponding to the micro wavelength range [1-3].

A mechanism for the formation of nanostructured states during SPD and the long-term operation of rails is proposed. It resides in the fact that tangential velocity discontinuity surfaces appear in the deformable material. The Kelvin-Helmholtz instability arises on these surfaces. An analysis of the dependence of the perturbation decrement on the wavenumber showed that this instability manifests itself both in the nanoscale and in the microsized wavelength range. By the example of rail steel after the long-term operation, it is shown that the critical wavelength ranges from 11 to 40 nm, which corresponds to the observed dimensions of structural elements.

A model is presented in [83] describing the mechanisms of crack initiation and material wear based on the data of metallographic studies in the wheel-rail contact zone. It is based on a multiscale description using the finite element method of deformation processes in surface layers upon contact. Comparison with experimental data on the surface topography is performed. The consequences of the model make it possible to predict the distribution of deformations in the range from millimeter to micron. The most complete description of models that take into account the variety of processes in the wheel-rail system is studied in monographs [1-3].

Given the complexity of the processes occurring in the metal of rails during the long-term operation, their interconnection, and mutual influence, the content and use of model concepts are very promising. Unfortunately, most of the models relate to the applied aspects of the problem and the formation of nanostructured phase states in the surface layers.

Model concepts of the processes occurring in surface nanolayers can be divided into two subgroups, considering the fundamental and applied aspects of the problem. Under severe plastic deformation of the surface during friction, a layer of material with a nanosized grain-subgrain structure is formed [84, 85].

In the resulting nanocrystalline material, individual crystallites practically do not contain dislocations, and it is deformed by the mechanism of grain boundary sliding [86]. The results obtained in [87] show that polycrystals deformed by this mechanism behave as if they had Newtonian viscosity. Consequently, it can be assumed that nanocrystalline layers flow like a viscous liquid with two flow regimes: laminar and/or turbulent. According to the authors of [86], the change from laminar to turbulent regime determines the moment of the onset of a seizure and catastrophic destruction of the contact surface. Analysis of the chronograms of the logarithm of the plastic shear rate [88] showed that the entire surface layer of the material under the action of the friction force moves along the sliding direction. Its movement can be compared with a laminar flow of a viscous fluid, the speed of which is not the same over the flow cross-section. Consequently, at different times at different depths, at the boundaries of elastic and plastically deformed regions, and within the zones of intense plastic shear, there are surfaces of tangential discontinuity of the velocity. From the standpoint of hydrodynamics, there is an absolute instability on such surfaces, which is the simplest case of Helmholtz instability, i.e. the absolute instability of a special type of interface separating from each other the flow regions filled with the same or different fluids moving at different speeds [89, 90].

A hierarchical multiscale method for analyzing the effect of lubrication on rolling contact fatigue was proposed in [91]. The method contains a molecular model of a lubricant and a continuous model of rolling contact components. At the nanoscale, molecular dynamics approaches have made it possible to estimate the coefficient of friction on the rolling surface. At the macro level, the finite element method made it possible to predict fatigue endurance based on the analysis of rolling contact components. The role of variable load on rolling contact fatigue has also been studied.

Finite element modeling allowed us to analyze the effect of structure gradients on the resistance to fatigue wear during sliding-rolling [92]. It was found that the gradient structures increase the resistance to fatigue wear due to the reduction of horizontal and vertical displacements on the rolling surface of the rails. With a fixed layer thickness having a gradient structure, an increase in surface strength properties (yield strength) can significantly reduce surface displacements. The implication of this paper may be of advantage for the production of premium rails.

The combined numerical model described in [93] allows predicting the service life by taking into account the abrasion of the rail ends. The model is based on Arhard's equation and Lemaitre's pitting model. The experimental evaluation carried out in laboratory conditions confirmed the adequacy of the proposed approach with a probability of 0.88.

Accumulation of plastic deformation is one of the causes of surface cracking. The situation is complicated by changing weather conditions that stimulate the growth of cracks. Changing the stable deformation regime is a key point in predicting the parameters of cracking [94]. The proposed model based on the integral equation allows predicting the stress field and crack morphology under a large loading cycle. The model potential also allows one to describe the cyclic plasticity of the material and predict the maximum expected depth of cracks in the wheels.

The dissolution process of cementite is usually analyzed by direct methods of modern physical materials science. In this regard, the results of [95], where the experimental data are compared with the phase-field model, are of undoubted interest. Three stages of cementite dissolution were established, confirming the gradient nature of the process in depth. The effects of the thickness of the ferrite/cementite interphase layers and the thickness of the cementite plates on the kinetics of the process were studied. The phase-field model developed by the authors allows us to take a fresh look at the predictions of mechanical properties and contact fatigue during the rolling of rails in operation.

The state of the working surface of the rails during operation is constantly monitored. On the experimental ring of JSC VNIIZhT, a comparative analysis of the change in the dimensions of the rail head after the operation is carried out according to the main controlled indicators: the correspondence of the actual dimensions of the new rails to the nominal ones, the intensity of the formation of vertical and lateral wear, the change in longitudinal vertical undulating irregularities, the formation of cracks and spalling on the rolling surface at the different amount of the hauled tonnage [96].

An analysis of operational data concerning the failure of rails is presented in [97], which showed that from year to year the most common damages are defects of a contact-fatigue nature. In connection with the development of heavy traffic on the railways of the Russian Federation, the specialists of JSC VNIKTI have developed a model for estimating the service life of rails before the formation of a contact fatigue crack on the rolling surface, depending on the values of the axial load. To determine the accumulation of contact fatigue damage on the rolling surface of the rails, a multiaxial fatigue model was selected. Taking into account the variability of the vertical load, the spectra of vertical forces were investigated, obtained by running tests on the impact on the track of freight trains formed from innovative 12-9853 cars with 18-9855 bogies and an axle load

Structure and Properties of Lengthy Rails after Extreme Long-Term Operation
Materials Research Foundations **106** (2021)

Materials Research Forum LLC
https://doi.org/10.21741/9781644901472

of 25 tf, 12-9548-01 cars on 18- 6863 with axle loads of 27 tf and serial cars on bogies 18-100 with axle loads of 23.5 tf. [97].

Kogan [98] presented a mathematical model of the emergence and development of wavy rail wear. The problem of assessing the vertical wear of the railhead under the axles of electric locomotives passing in the traction mode is considered. A technique has been developed that makes it possible to build a chain of calculations that determine the occurrence and development of undulating wear of the railhead. A specific example of the calculation is given, illustrating the process of propagation of undulating wear from the source of its occurrence in the direction of train movement.

The work [99] considers in detail the approach that uses the criterion of contact fatigue (Dang Wang criterion) and the diagram of adaptability of materials to the action of alternating stresses (Johnson diagram). This approach has found wide application to assess the contact fatigue of wheels of railway rolling stock and rails. A criterion based on the maximum value of shear stress is also considered.

Welded rail joints, joined by methods of electric contact welding in stationary and field conditions, as well as by methods of aluminothermic welding, are weak points of the track and are damaged more often than rails outside the zone of welded joints.

Shur [100] showed ways of eliminating increased damage to rails in the zone of welded joints. At present, the weak point of welded joints in electrical contact welding is not only defects associated with lack of penetration in the welding zone, burns in places of poor contact of the rail foot with current-carrying contact jaws, or defects in the machining of seams after welding, but also a local decrease in hardness in heat-affected zones after welding and local heat treatment. In this case, the elimination of wide zones of low hardness in the places of welded joints, leading to the formation of saddles, cracks and chipping (defects 46.3 and 16.3), and contributing to the destruction of rails near such joints by defects 75.2, 79.2 and 21.2, becomes the main task of increasing the efficiency of welded joints [100].

The nature of friction and wear is analyzed in [101, 102]. Two theories of friction are compared: the generally recognized adhesion-deformation and alternative dehision-deformation, the foundations of which were laid by Prandtl and Deryagin. The concepts and effects of the action of adhesive and cohesive forces of interatomic attraction are clarified. Concepts and peculiarities of the effect of the interatomic repulsion dehision forces are specified. The dehision-deformation theory was developed and its main provisions were formulated. When two atoms approach each other, an adhesive attraction arises, which then transforms into a cohesive bond, and then into a dehision repulsion (Lennard-Jones dependence). The adhesive attraction also occurs between substances containing polarized molecules (surfactants). When solids are compressed, the atoms of one body are pressed into the interatomic gaps of another, forming potential barriers - atomic-molecular roughness, which creates resistance to the tangential displacement of bodies. At the same time, repulsive forces arise between the atoms, as between the like poles of a magnet, which, having reached the point of unstable equilibrium (bifurcation point), can instantly be replaced by the forces of cohesive attraction. During friction, the inversion (change) of forces with the formation of new compounds is rarely observed, since the bifurcation pressure (branching of processes) in most pairs of substances is above the plasticity limit. Friction forces of solids arise as a result of the engagement of atomic-molecular roughness irregularities on the spots of actual contact. The proportionality of the friction force to the normal force (Coulomb's law) is ensured, first, by a proportional increase in the height of the potential barriers of atomic roughness, and, second, by an increase in the area of actual contact spots due to plastic or elastic deformation. The real pressure under conditions of volumetric compression in the zones of actual contact of solids, even at ultra-low loads, is so high that during shear, there is not a jump of particles from one potential depression to another, but a shear of surface sections. The cut-off wear particles behave like solids, are pressed into hard conglomerates and scratch surfaces. The dry friction force is determined by the contact pressure, the strength of the conglomerates, which are fragments of the interlayer, and the strength of the bonds of the fragments of the interlayer with each other and with the friction surfaces. During fluid friction, molecules move in a barrier-free, but the non-uniform field of cohesive forces, which determines the absence of static friction and the dependence of friction forces on the flow velocity (Newton's law), in contrast to friction in gases resulting from the collision of freely moving molecules (Maxwell's law). Liquids that do not contain polarized molecules are squeezed out of the friction zone and do not affect the friction force. If the lubricating fluid contains surfactants, then the friction force can vary from the maximum solid-state to the minimum liquid depending on the viscosity, surfactant content, pressure, contact area, sliding, and rolling speeds (Stribeck dependence) [101].

Contact fatigue damage to rails, along with their wear, are the most common types of rail defects, including new-generation rails. In recent years, there have been significant changes in the distribution of types of contact-fatigue damage both on Russian and foreign railways, especially on lines with heavy traffic. Therefore, it is relevant to study the mechanisms and simulate the appearance of surface contact-fatigue damage to rails. A review paper [102] is devoted to approaches to modeling the occurrence of contact-fatigue damage on the working surfaces of rails. Four types of approaches to modeling are considered: based on the methods and approaches of the mechanics of contact interaction; on obtaining quantitative characteristics of the adaptability of materials to cyclic loading, established using laboratory tests; on the application of criteria that have the physical meaning of the energy released at the contact; predicting the accumulation of plastic deformation under cyclic loading based on a series of standard tests of rail steels, including in the welded joint zone, and finite element modeling. Future research on the formation and development of surface contact-fatigue defects in rails is also formulated [102].

Conclusion

The analysis of the literature given in [1-3] shows that an increase in the intensity of railway traffic and its load density necessitates a further increase in the operational resistance of rails. The service life of rails depends on various factors: steel cleanliness, railhead hardening technology, lubrication, operating conditions, etc. The formation of numerous defects during the operation is one of the main reasons for the failure of rails. Therefore, the issues of resistance to the accumulation of damage and, first of all, wear and fatigue, are the subject of the most careful study from the standpoint of both scientific research and experimental design, and technological developments.

The problem of the formation and evolution of the structure as well as properties of rails during long-term operation is a complex set of interrelated scientific and technical issues. One of the most important directions contributing to the development of ideas about the nature of structural-phase transformations is the establishment of the corresponding quantitative regularities over the cross-section of rails. Significant experimental difficulties associated with the use of methods of modern physical materials science and, above all, transmission electron diffraction microscopy, have led to a relatively small number of studies in this area, completely incommensurate with the number of works on smelting rail steel, rolling and calibration of rails, optical microscopy of the rail structure, model representations, etc.

Taking into account that the kinetics of the processes of formation of structural-phase states is related to the foundations of the theory of strength and plasticity, it is extremely important to have information on the parameters of the fine structure of rails in their different sections. In this respect, the results of the analysis of dislocation substructures and extinction contours, which make it possible to estimate the level of internal long-range stress fields, may turn out to be interesting. The formation of a structure with nanosized particles in the surface layers of rails after severe plastic deformation has been noted in a limited number of publications. There are practically no works on the quantitative analysis of various contributions (lattice friction, intraphase boundaries, dislocation substructure, particles of the carbide phase, long-range stress fields) to the strengthening of rails. The obvious scientific fundamental significance of the directions of these studies is combined with the fact that understanding the physical nature and the main parameters of the formation and evolution of structural-phase states and dislocation substructure is already desirable today, and tomorrow, as the requirements for the mechanical and operational properties of rails grow, it is an exceedingly necessary condition for obtaining high-quality rails with high performance.

Summing up the analysis of the available literature, it should be noted that the establishment of the physical mechanisms of the formation and evolution of structural-phase states and dislocation substructure in rails during the operation is one of the most important tasks of condensed matter physics, metal science, and heat treatment.

Thus, it is quite clear that it is of scientific and practical interest to study the patterns and nature of the formation of the structure, phase composition, and dislocation substructure of rail steel at different periods and operating conditions. This work is devoted to the solution to these problems. A comparative analysis of the results obtained by the methods of modern physical materials science (first of all, by employing transmission electron microscopy) was performed on the structure, phase composition, and defect structure, formed along the central axis and the fillet in the head of differentially hardened 100-meter rails after various periods of operation.

2. Material and Methods of Research

2.1 Research material

The studies were carried out on differentially hardened rails of the R350HT category made of steel grade E76HF in the initial state and after the hauled load of 691.8 and 1411 million gross tons during the field tests at the Experimental Ring of JSC VNIIZhT.

The results of the chemical analysis of the rail metal composition show that in terms of the content of all chemical elements, the rail metal meets the requirements of GOST R 51685-2013[1] for steel grade E76HF. This steel grade is subject to a regulatory document EN 13674-1:2011+A1:2017, and its mass fraction of chemical elements of the steel is given below in Table 2.1.

Table 2.1 *Mass fraction of chemical elements (%) in rails R350HT.*

C	Mn	Si	V	Cr	P	S	Al
0.70-0.82	0.65-1.25	0.13-0.60	Max 0.030	≤ 0.15	Max 0.025	Max 0.030	Max 0.004

2.2 Methods for studying the structure

Analysis of the steel structure is carried out on the rolling surface (along the central axis) and fillets (along the radius of rounding). The term "fillet" denotes the convex surface of the railhead, described by the radius of the rounding, and connecting the rolling surface and the upper part of the side edge of the head (GOST R 50542-93) [103].

The structure, elemental and phase composition, and defect substructure were investigated by optical microscopy, X-ray diffraction analysis, and transmission micro-diffraction electron microscopy. The chemical composition was determined by X-ray spectral analysis on a Shimadzu XRF-1800 sequential X-ray fluorescence wave-dispersive spectrometer.

The quantitative analysis of the steel structure was carried out using the methods of stereology [104, 105] and quantitative electron microscopy [106, 107]; phase analysis of steel was carried out by indexing electron diffraction patterns using the dark-field technique [108-113].

---

[1] Hereinafter, the abbreviation GOST refers to All-Union State Standard adopted in the former USSR and still used in some post-Soviet republics.

2.2.1   Macro and microstructural studies

Metallographic studies were carried out at the Center for Collective Use "Materials Science" of the Siberian State Industrial University.

The method for preparing samples for optical studies did not differ from the generally accepted one and included: cutting out samples, pouring into a mold, grinding on sanding paper of varying grain size, polishing with polishing (diamond) pastes, and etching.

Metallographic studies were performed on an Olympus GX51 optical microscope equipped with a digital camera with Siams Photolab 700 software.

The macrostructure of the rail metal was determined by the requirements of GOST R 51685-2013 on a full-profile template cut from the rail in the transverse direction after etching in a 50% aqueous solution of hydrochloric acid. The study of the structure by optical microscopy methods was carried out on non-etched thin sections and after electrolytic polishing of the surface in a 5% acetic solution of perchloric acid followed by etching in a 4% alcohol solution of nitric acid [114].

Inadmissible defects of the macrostructure such as flocs, delamination, cracks, crusts, spotted liquation, foreign metal, and slag inclusions lead to material rejection. The assessment of metal contamination with non-metallic inclusions was determined according to GOST R 51685-2013. Following GOST 1778-70, the samples were assessed for the presence of sulfides, point oxides, and brittle silicates.

Analysis and evaluation of the microstructure of the samples and the degree of structural components were carried out per GOST 8233.

Determination of the average grain size was carried out by the method of random secants on microsections. The grain boundaries were etched electrolytically in a reagent. The average grain size (D) in the volume of the material was determined on the basis of the average grain size measured by microsections [104]:

$$\bar{D} = 0.5\pi(d^{-1})^{-1} \tag{2.1}$$

where $\bar{d}$ is the average grain size determined by the microsection:

$$\bar{d}^{-1} = N^{-1} \sum_{i=1}^{N} d_i^{-1} \tag{2.2}$$

where N is the number of measurements, d_i is the current grain size on the microsection. The standard deviation (σ_D) was determined using the following expression:

$$\sigma_D = \sqrt{\frac{4}{\pi(\bar{d} \cdot \bar{D})} - (\bar{D})^2} \tag{2.3}$$

2.2.2   X-ray structural studies

The phase composition and state of the crystal lattice were studied by X-ray diffraction analysis (XRD-6000 diffractometer, Shimadzu) [115].

X-ray studies carried out in the work consisted in determining the lattice parameter of the α-phase; the size of the regions of coherent scattering and the magnitude of microstresses. The study was carried out using CuKα radiation (wavelength λCu = 1.540598 Å). Diffraction patterns were obtained with Bragg-Brentano focusing in the ($\theta - 2\theta$) scanning mode (θ is the angle between the incident beam and the reflecting atomic plane) in the angle range 2θ from 10° to 120° with a step of 0.02°. The phase composition was analyzed using the PCPDFIN databases and the POWDER CELL 2.5 full-profile analysis program.

The calculation of the cubic lattice parameter (a_0) was performed using the relation [112]:

$$a_0 = d_{hkl}\sqrt{h^2 + k^2 + l^2}$$

where d_{hkl} is the interplanar distance; h, k, l - plane indices.

The sizes of coherent scattering regions (CSRs) were calculated at small diffraction angles using the Scherrer formula [115]:

$$D_{hkl} = \frac{K \cdot \lambda}{\beta \cdot cos(\theta_{hkl})}$$

where D_{hkl} is the CSR size, K is the coefficient taking into account the shape of the particles, λ is the X-ray wavelength, β is the half-width of the X-ray reflection, and θ_{hkl} is the diffraction angle. The calculation was carried out at $K = 0.94$.

Microdistortions of the b-iron crystal lattice ($\Delta d/d$) were calculated from the broadening of the reflection using the expression [115]:

$$\frac{\Delta d}{d} = 0.25\beta \cdot ctg(\theta_{hkl})$$

The revealed quantitative regularities of changes in the parameters of the rolling surface structure along the central axis of the rail steel after the hauled load of 691.8 million gross tons made it possible to carry out research aimed at analyzing the distribution of carbon atoms in the steel structure.

2.2.3   Electron microscopic examination

The study of the defective substructure, phase morphology, and the state of the carbide phase of the rails after the operation was carried out using transmission diffraction electron microscopy (using an EM-125 microscope). Foils for research were made by the methods of electrolytic thinning of plates cut by the electric spark method at a distance of

2 mm, 10 mm and near the rolling surface along the central axis (Fig. 2.1, a) or a working fillet (Fig. 2.1, b). The sample preparation scheme is shown in Fig. 2.1.

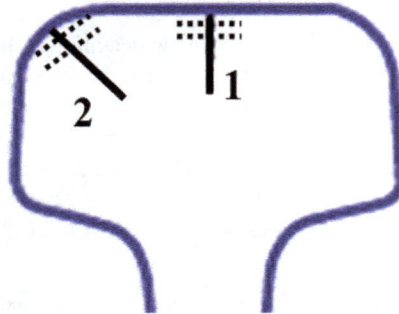

Figure 2.1 *Schematic drawing of preparation of a rail sample in the study of its structure by methods of optical and electron diffraction microscopy. Solid lines show directions along the central axis (1) and along the fillet (2); the dotted lines conventionally indicate the locations of the metal layers used to prepare the foils (surface, 2, 10 mm from the surface).*

Bright-field images of the fine structure were used to classify morphological features of the structure, to determine the size, volume fraction, and locations of phases. The following technique was used to identify new phases. Having obtained a bright-field image of the structure of the material, the particles of the second phase of interest were studied. For this, at least three dark-field images were obtained in different reflections of these particles, and reflections that belonged to the crystal lattice of these particles were identified. After photographing the microelectron diffraction pattern from the analyzed portion of the foil, the radius vectors of the reflections belonging to the analyzed particle and the angles between them were measured. Next, a reciprocal lattice was constructed containing the identified reflections of the analyzed particle. After that, phase identification was reduced to finding such a reciprocal lattice, the cross-section of which could be represented by a given point electron diffraction pattern.

In the case of isotropic structure, P_V can be determined on one random cross-section of the crystal. For a heterogeneous structure, a representative sampling must be carried out over several differently oriented sections.

The scalar dislocation density was measured by the secant method with a correction for the invisibility of dislocations [109, 111, 112]. A rectangular mesh was used as a test line.

Then the scalar dislocation density in micrographs obtained by electron microscopic examination can be determined as follows

$$\rho = \frac{M}{t}\left(\frac{n_1}{l_1} + \frac{n_2}{l_2}\right) \tag{2.4}$$

where M is the magnification of the micrograph, n_1 and n_2 are the numbers of intersections of horizontal l_1 and vertical l_2 lines by dislocations, respectively l_1 and l_2 are the total lengths of horizontal and vertical lines.

The scalar dislocation density was determined separately for each type of dislocation substructure (DSS). The average value of the scalar dislocation density was calculated taking into account the volume fraction of each type of the present DSS using the following formula:

$$\rho = \sum_{i=1}^{Z} P_{Vi}\rho_i \tag{2.5}$$

where ρ_i is the scalar density of dislocations in a certain type of DSS, P_{Vi} is the volume fraction of the material occupied by this type of DSS, Z is the number of types of DSS.

The excess dislocation density $\rho_\pm = \rho_+ + \rho_-$ (ρ_+ and ρ_- are the density of positively and negatively charged dislocations, respectively) was measured locally using the misorientation gradient [116-119]:

$$\rho_\pm = \frac{1}{b}\frac{\partial \varphi}{\partial \ell} \tag{2.6}$$

where b is the Burgers vector of dislocations, $\partial\varphi/\partial\ell$ is the gradient of the foil curvature or the curvature-torsion of the crystal lattice χ. The value $\chi = \partial\varphi/\partial\ell$ was determined by shifting the extinction contour ($\Delta\ell$) at a controlled angle of inclination of the foil ($\Delta\varphi$) in the microscope column using a goniometer. In this case, the vector of the effective reflection $\vec{g}$ should be perpendicular to the axis of inclination of the goniometer (AIG). Otherwise, recalculation is required, since the plane of the effective reflection will no longer contain the tilt axis of the goniometer. It should be noted that the section of the foil on which the measurement is carried out should not contain interfaces or misorientations on the path of the contour displacement, i.e. the bending of the foil must be continuous. It was established by special experiments that the width of the contour in terms of misorientations for fcc alloys based on nickel, copper and iron [116], and martensitic steels [119, 120] is approximately one degree. This means that when the goniometer is turned by an amount $\Delta\varphi \approx 1°$, the bending extinction contour is displaced by a distance equal to its width, i.e. $\Delta\ell \approx \ell$ (in this case, the condition $\vec{g} \perp$ AIG must be satisfied).

This value ($\Delta\varphi \approx 1°$), in combination with the contour width ℓ, makes it possible to determine the misorientation gradient:

$$\frac{\partial\varphi}{\partial\ell} = 1.7 \times 10^6 \cdot \frac{1}{\ell} \text{ [rad/cm]} \tag{2.7}$$

The magnitude of the internal stress fields, following the calculation scheme proposed by E.V. Kozlov and N.A. Koneva can be estimated as follows when applied to the steel [121]:

$$\sigma = G \cdot t \cdot \chi$$

Here G is the shear modulus of the material under study; t is the thickness of the foil (for the EM-125 electron microscope, t $\approx$ 200 nm); Here $\chi = \frac{\partial\varphi}{\partial\ell}$ is the curvature-torsion of the crystal lattice, where $\partial\varphi$ is the angle of change in the orientation of the reflecting plane of the foil; $\partial\ell$ is the amount of movement of the bending contour.

In the study of structural steels, it was shown [120, 121] that the width of the contour h in terms of misorientations is approximately one degree (about 0.0175 rad.). Therefore,

$$\chi = \frac{\partial\varphi}{\partial\ell} = \frac{0.0175}{h}$$

Hence,

$$\sigma = G \cdot t \cdot \chi = G \cdot t \cdot \frac{0.0175}{h} = G \cdot \frac{3.5}{h} \tag{2.8}$$

When making estimates, the width of the contour h should be measured in nanometers.

2.3 Methods for determining mechanical properties

The hardness of the rail metal was determined on a transverse template cut out and prepared following the requirements of GOST R51685-2013. The hardness test was carried out by the Brinell method on a TSh-2M hardness tester with a ball 10 mm in diameter at a load of 3000 kgf under the requirements of GOST 9012-59.

Tribological tests of rail steel (tests were carried out for two sections of the rails, namely, on the rolling surface and at a distance of 15 mm along the central axis from the rolling surface) were carried out on a Pin on Disc and Oscillating TRIBOtester (TRIBOtechnic, France) with the following parameters: a ball made of ShKh15 steel with a diameter of 6 mm, track radius 2 mm, load 10 N, distance 30-80 m. Measurements were carried out according to the scheme "sample rotation with a fixed counter-body"; linear rotation speed (2.0 - 2.5) cm / s. The test scheme is shown in Fig. 2.2. At the end of the friction process, the tribometer was used to measure the profile of the friction groove on the

Structure and Properties of Lengthy Rails after Extreme Long-Term Operation
Materials Research Foundations **106** (2021)

Materials Research Forum LLC
https://doi.org/10.21741/9781644901472

Tensile mechanical properties were determined on a tensile testing machine EU-40 with a force of 10 tons on two tensile cylindrical specimens with a diameter of 6 mm and an initial design length of the working part of 30 mm, prepared per the requirements of GOST R 51685-2013 and GOST 1497.

surface of the samples with a numerical analysis of the depth of the friction groove and its cross-sectional area. Wear resistance was assessed as the reciprocal of the wear or intensity rate [122]. The wear rate was calculated using the following formula:

$$V = (2\pi \cdot R \cdot A)/(F \cdot L),$$

where R is the radius of the track [mm], A is the cross-sectional area of the wear groove (mm^2), F is the value of the applied load (N), L is the distance traveled by the counter-body ball (m).

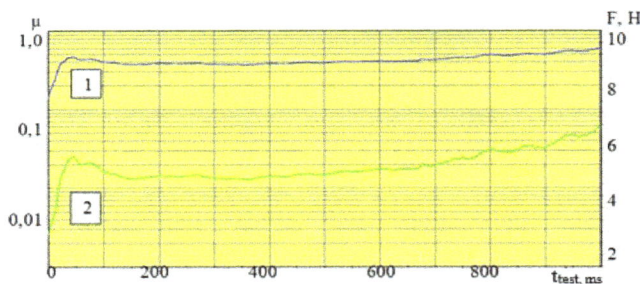

Figure 2.2 *Scheme and conditions of tests for wear resistance of rail metal (curve 1 - coefficient of friction (μ), curve 2 - friction force (F)).*

A typical image of the rail metal friction groove profile is shown in Fig. 2.3.

Impact bending test was carried out on an MK-15 pendulum tester following the requirements of GOST 9454 on standard specimens $10 \times 10 \times 55$ mm in size with a *U*-shaped notch with a radius of one millimeter and a depth of two millimeters at temperatures of plus 20 ^{0}C and minus 60 ^{0}C (optional).

Research methods are described in more detail in monographs [1-3, 123].

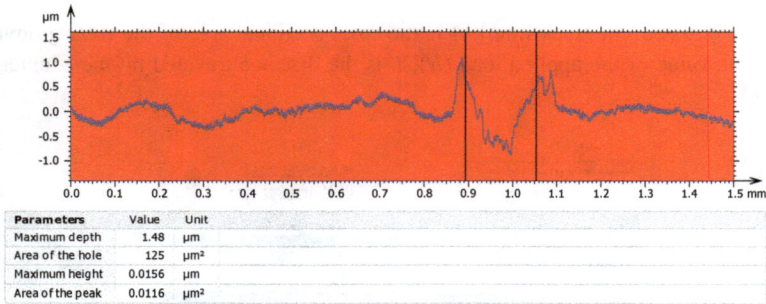

Parameters	Value	Unit
Maximum depth	1.48	μm
Area of the hole	125	μm²
Maximum height	0.0156	μm
Area of the peak	0.0116	μm²

Figure 2.3 *Profile of the rail metal friction groove formed during tribological tests.*

3. Differentially Hardened Rails: Structure, Phase Composition, and Dislocation Substructure

Studies carried out by metallography methods [1-3, 123-140] have shown that the microstructure of the surface layer of steel, regardless of the direction of analysis, is predominantly fine-dispersed thin-lamellar pearlite. Along the boundaries of pearlite grains, structurally free ferrite is discovered in the form of a grid (near the surface in a layer ~ 150 μm thick), and in the form of separate scattered inclusions (at a distance from the sample surface up to ~ 300 μm). The actual grain size, assessed by the continuous ferrite mesh along the grain boundaries, varies within 6-7 numbers of the GOST 5639 scale. At a depth of 10 mm, the steel microstructure is represented by pearlite. Ferrite is observed in the form of inclusions sparsely located along the grain boundaries. Typical images of the structure of an etched section, which demonstrate a change in the structure of steel depending on the distance to the cooling surface, are shown in Fig. 3.1.

Thus, differentiated quenching is accompanied by the formation of a pearlite structure interspersed with grains of structurally free ferrite along the boundaries. With increasing distance from the cooling surface, the relative content of ferrite decreases [1-3, 123-140].

The results of the X-ray phase study of the samples are given in Table. 3.1.

Table 3.1 Results of X-ray structural analysis of rails subjected to differential hardening.

Direction of analysis	Phase composition, %		a (Fe), nm	$\Delta d/d$	D_{csr}, nm
	Fe	Fe$_3$C			
axis	89	11	0.28638	0.005	108.9
fillet	92.7	7.3	0.28664	0.007	83.7

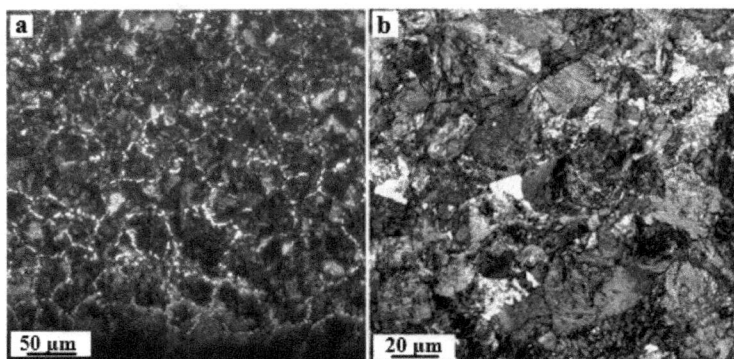

Figure 3.1 Microstructure of rails at the surface (a) and a depth of 10 mm (b) from the rolling surface [1-3, 123].

By analyzing the results given in Table. 3.1, it is possible to note differences in the state of steel along the central axis and along the fillet. Namely, in the volume of the material on the central axis (concerning the volume of the material along the fillet), the relative content of cementite is higher, the lattice parameter of α-Fe is lower, the microstresses ($\Delta d/d$) are lower, the sizes of the coherent scattering regions (D_{csr}) are larger than those of the volume of steel located on the fillet. Taken together, these facts indicate a higher cooling rate of the volume of the material located on the fillet to the volume of the material located along the central axis [1-3, 123-140].

It has been established by transmission electron microscopy of thin foils that the structure of the steel, regardless of the distance of the studied layer to the rolling surface, is represented by pearlite grains of plate morphology (Fig.3.2, a), grains of structurally free ferrite (ferrite grains that do not contain carbide particles in the bulk phases) (Fig. 3.2, b) and ferrite grains, in the volume of which cementite particles are observed[2] mainly in the form of short plates (Fig. 3.2, d), and globular particles (Fig. 3.2, c). As a rule, the volumes of steel with globular particles and particles in the form of short plates are observed separately, which made it possible to estimate their relative content in the material, equal to 1: 10.

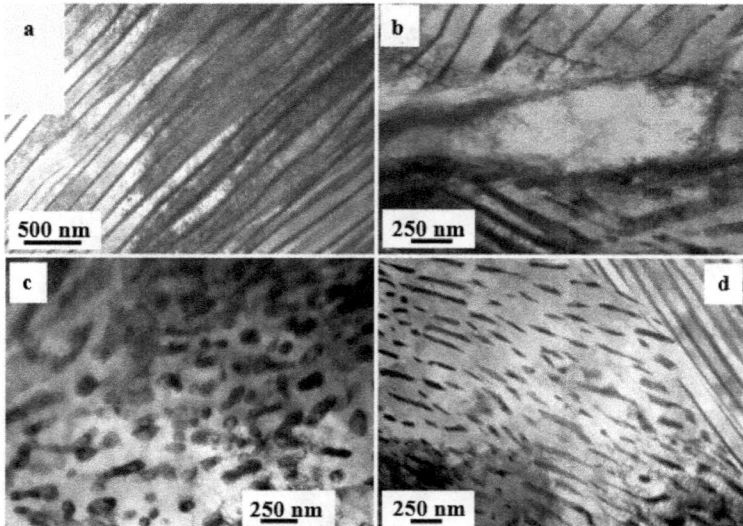

Figure 3.2 *TEM images of the rail structure [1-3, 123].*

---

[2] Hereinafter, grains of a ferrite-carbide mixture

The relative content of the identified types of the structure depends on the position of the depth of the studied layer (Table 3.2). Analyzing the results shown in this table, it can be noted that the relative volume fraction of grains of structurally free ferrite V(3) is small and varies from 0.01 to 0.05 of the steel structure. The relative volume fraction of grains of the ferrite-carbide mixture V(2) is significantly weightier, the value of which varies in the range from 0.17 to 0.37 of the steel structure and decreases with distance from the rolling surface [1-3, 123-140].

Table 3.2 *The relative content of the structural components of the R350HT rails.*

Distance from the surface, mm	Central axis			Fillet		
	V(1)	V(2)	V(3)	V(1)	V(2)	V(3)
0	0.67	0.28	0.05	0.61	0.37	0.02
2	0.82	0.17	< 0,01	0.63	0.34	0.03
10	0.73	0.26	< 0,01	0.71	0.28	< 0,01

Some regularity can be seen in the change in the structure of steel depending on the location of the analyzed layer (on the central axis or the fillet). Namely, regardless of the distance to the rolling surface on the fillet, the proportion of grains of lamellar pearlite is lower, and the grains of the ferrite-carbide mixture is higher than on the central axis. At a depth of about 10 mm, these differences in the steel structure become the same. Consequently, the processes of phase transformations that take place during the differentiated quenching proceed according to slightly different thermokinetic diagrams relative to the volume of steel along the central axis and on the fillet.

The characteristic of pearlite, regulated by the GOST, is the value of the inter-plate distance. The results of the performed estimates show that the average value of the interplate distance varies from 120 to 190 nm and decreases with distance from the rolling surface both along the central axis and along the fillet. Following GOST 8233-56, we can say that the pearlite structure of the surface layer (approximately 10 mm thick) of the investigated rail steel belongs to the first point, is characterized as sorbitic, the type of structure is troostite [1-3, 123-140].

The ferritic component of the steel structure (grains of structurally free ferrite, the ferrite component of pearlite grains, and grains of a ferrite-carbide mixture) is defective. A dislocation substructure in the form of chaotically distributed dislocations (Fig. 3.3, a) and a network (Fig. 3.3, b) dislocation substructure were discovered through electron microscopy. The scalar dislocation density in the steel studied varies from 4×10^{10} to 5.5×10^{10} cm^{-2}. In this case, the value of the scalar dislocation density in the grains of the ferrite-carbide mixture is somewhat higher than in the ferrite component of the pearlite grains, regardless of the location (along the central axis or on the fillet) and the distance of the studied layer from the rolling surface.

A feature of the fillet structure is the presence in a layer of about 2 mm thick of nanosized (5 ... 10 nm) particles of the carbide phase, which are found exclusively in ferrite plates of pearlite colonies (Fig. 3.4, a). This fact indicates the repeated decomposition of the solid solution based on α-iron, which takes place after the formation of the pearlite structure. The absence of such nanosized particles in the steel structure on the central axis is evidence in favor of the above assumption about a higher cooling rate of the surface layer of the fillet.

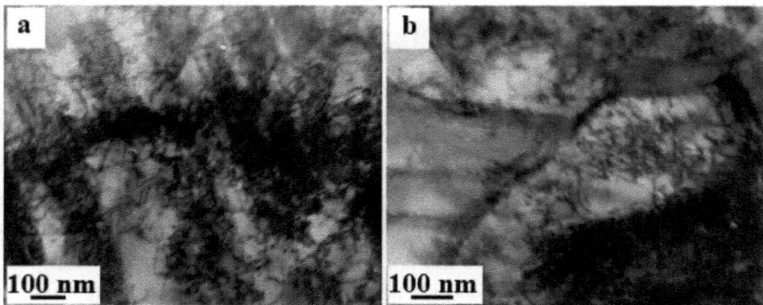

Figure 3.3 Electron microscopic image of the dislocation substructure [1-3, 123].

Another interesting fact discovered when analyzing the structure of the surface layer (approximately 2 mm thick layer) of steel along the fillet is the presence of speckled contrast in the image of cementite plates of pearlite colonies (Fig. 3.4, b). The presence of such a contrast indicates the defectiveness of cementite plates, which may also indicate a sufficiently high cooling rate of the volume of the surface layer of steel on the fillet.

Thus, the analysis of the results presented in this section, obtained in the study of the phase composition and defect substructure of differentially hardened R350HT rails, shows that the differentiated hardening of steel is accompanied by the formation of a morphologically multifaceted structure, represented by grains of lamellar pearlite, grains of a ferrite-carbide mixture, and grains of structure-free ferrite, located in the form of inclusions along the grain boundaries of pearlite [1-3, 123-140]. With an increase in the distance from the cooling surface, the relative content of grains of structurally free ferrite decreases. The facts revealed as a result of the studies performed indicate a higher cooling rate of the volume of the material located on the fillet in relation to the volume of the material located along the central axis.

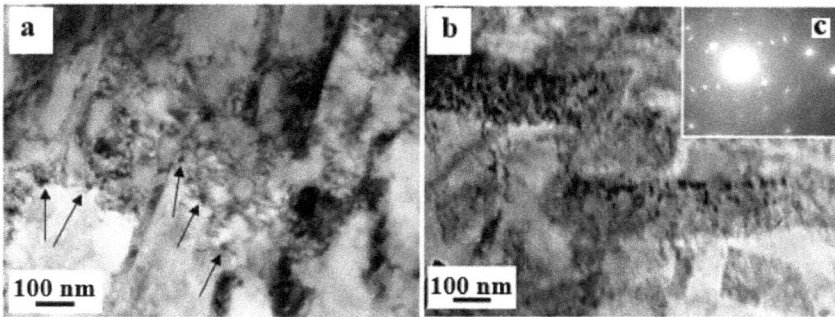

Figure 3.4 *TEM images of nanosized particles of the carbide phase; (a, b) bright-field images; (c) micro-electronogram; in (a), arrows indicate particles of the carbide phase [1-3, 123].*

It is obvious that the strength characteristics of rail steel, like that of any other material, are determined not only by the ratio of its structural components but also by the state of its defective substructure. In this regard, let us consider the substructure of the identified components of the rail steel structure in more detail.

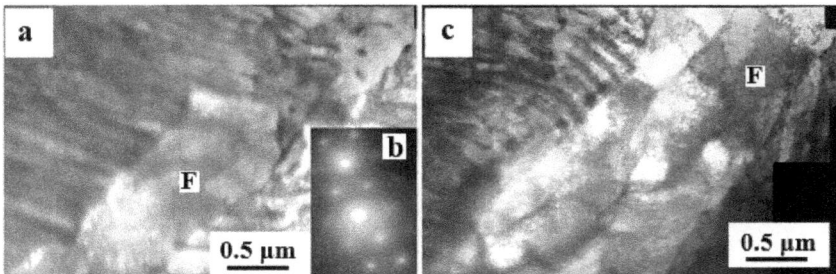

Figure 3.5 *TEM images of grains of structurally free rail ferrite; (a, c) bright-field images; (b) micro-electronogram; the symbol "F" denotes a grain of structurally free ferrite [1-3, 123].*

Grains of structurally free ferrite are arranged in chains or extended interlayers between pearlite grains (Fig. 3.5). The grain sizes of structurally free ferrite vary from tenths to a few micrometers. Often, cementite particles of a predominantly globular shape are located along the boundaries of such grains. Particle sizes vary from tens to hundreds of nanometers.

Lamellar pearlite grains are mostly imperfect. Often the cementite plates are bent and not parallel to each other, have different types of intergrowths, and ferrite bridges are observed (ferrite areas dividing the cementite plate) (Fig. 3.6).

The dispersion of the pearlite structure was assessed by the value of the interplate distance. The results of these evaluations are shown in Table. 3.3 and Fig. 3.7.

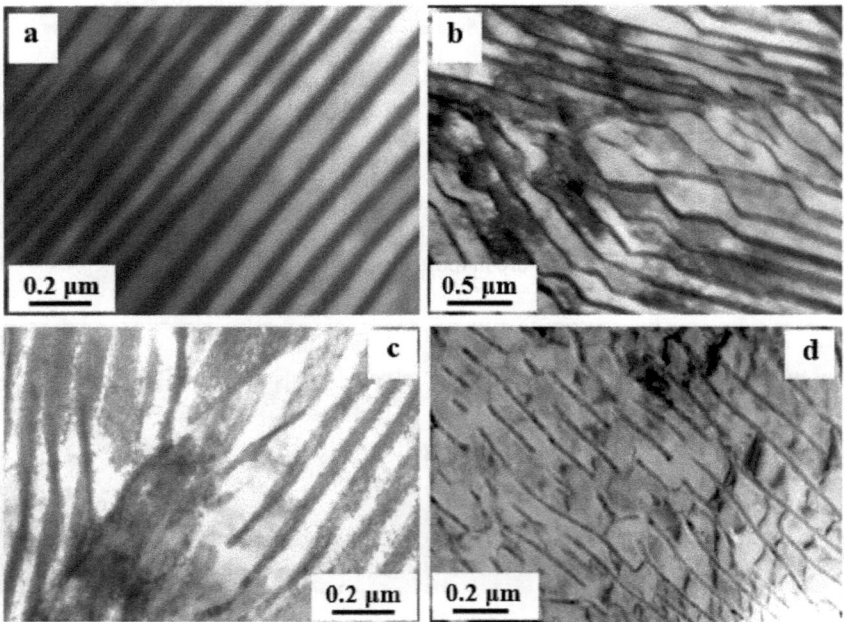

Figure 3.6 TEM image of pearlite with lamellar rail morphology [1-3, 123].

Table 3.3 Average values of the inter-plate distance (h) of pearlite grains of rails.

Distance from the surface, mm	h, nm (central axis)	h, nm (fillet)
0	165	190
2	120	135
10	135	125

Analyzing the results presented in this way, it can be noted that the value of the interplate distance varies from 120 to 190 nm and decreases when passing from the rolling surface to the layer located at a depth of 10 nm [1-3, 123-140].

The interplate distance averaged over the investigated surface volume of steel of about 10 mm thick is 145 nm. Following GOST 8233-56, we can say that the pearlite structure of all samples of the investigated rail steel refers to the first point, is characterized as sorbitol, the type of structure is troostite.

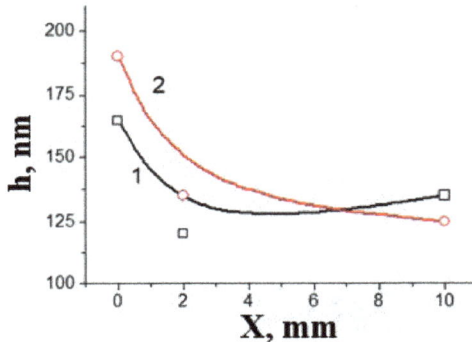

Figure 3.7 *Dependence of the inter-plate distance of lamellar pearlite grains on the distance from the rolling surface. (1) analysis along the central axis; (2) along the fillet [1-3, 123].*

The grains of the ferrite-carbide mixture are characterized by the presence of particles of the carbide phase in the volume, the size and shape of which vary within wide limits. Fig. 3.8 shows electron microscopic images of the most characteristic grains of the ferrite-carbide mixture, which differ in the shape and size of the particles of the carbide phase.

The grains of the ferrite-carbide mixture are fragmented, i.e. divided into areas with slight misorientation (Fig. 3.9). Lamellar or globular particles of the carbide phase are located along the boundaries of the fragments. Particle sizes vary from 20 to 40 nm. The average fragment size is 120 nm. The fragments are separated by low-angle boundaries. The value of the azimuthal component of the total angle of disorientation of fragments $\Delta\alpha$, determined by the method presented in [1-3], varies from 5.1 to 10.0 deg. (Fig. 3.10) (Table 3.4). Averaging over the analyzed volume of samples (volume of a layer 10 mm thick) shows that the value of the angle $\Delta\alpha$ for heat-strengthened rails is 6.3 degrees.

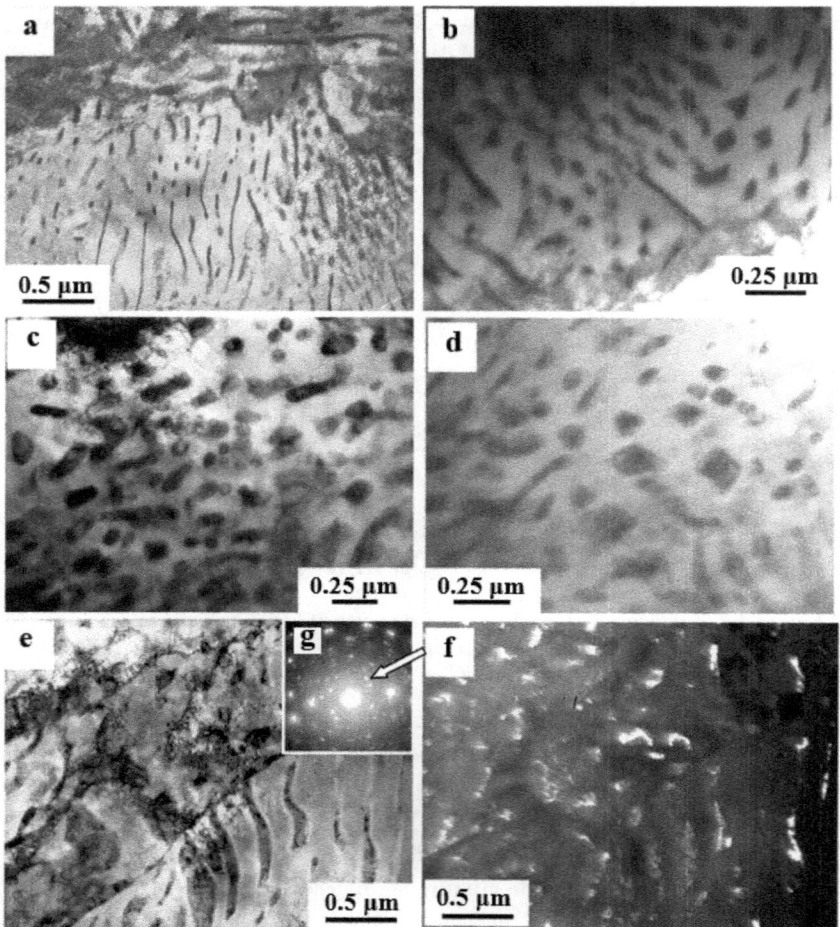

Figure 3.8 *TEM image of grains of a ferrite-carbide mixture; (a – e) bright-field images; (f) dark-field obtained in the reflection [122] Fe₃C; (g) – micro-electronogram (the arrow indicates the reflex in which the dark field was obtained) [1-3, 123].*

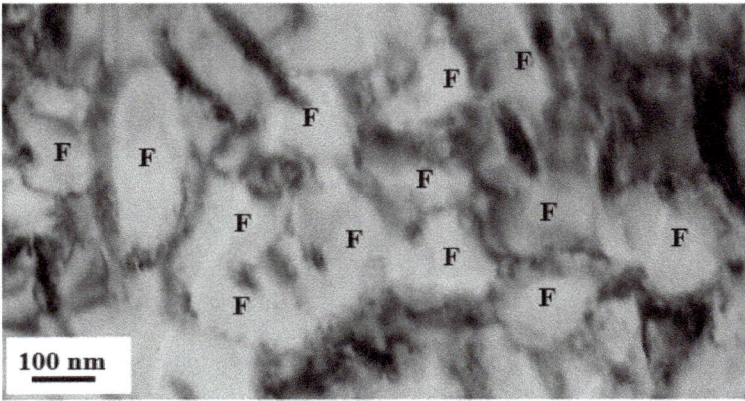

Figure 3.9 Substructure of grains of a ferrite-carbide mixture of rails. The symbol "F" denotes fragments [1-3, 123].

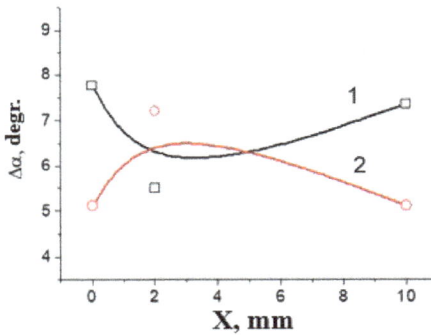

Figure 3.10 Dependence of the average value of the azimuthal component of the total angle of misorientation of fragments of grains of the ferrite-carbide mixture on the distance from the rolling surface. (1) analysis along the central axis; (2) along the fillet [1-3, 123].

As noted above, the grains of the ferrite-carbide mixture are characterized by the presence of particles of the carbide phase in the volume of the ferrite grain. Particles of the carbide phase vary in size and shape. Particles of a globular shape and particles in the form of short plates are distinguished [1-3, 123-140].

Table 3.4 *Average value of the azimuthal component of the total angle of misorientation of fragments of grains of a ferrite-carbide mixture of rails.*

Distance from the surface, mm	$\Delta\alpha$, degrees (central axis)	$\Delta\alpha$, degrees (fillet)
0	7.75	5.1
2	5.5	7.2
10	7.35	5.1

From the analysis of characteristic electron microscopic images (Fig. 3.8) of the particles of the carbide phase present in the volume of grains of the ferrite-carbide mixture, it can be assumed that the grains with globular particles of cementite are grains of globular pearlite; grains with cementite particles in the form of short plates can be attributed to grains of highly defective lamellar pearlite.

Analyzing the results presented in Table. 3.5 and Fig. 3.11, it can be noted that the volume fraction of grains with globular particles of the carbide phase increases with distance from the sample surface. Averaging over the analyzed volume of samples (volume of a layer 10 mm thick) shows that the volume fraction of grains with globular particles of the carbide phase for rails is 4.2%.

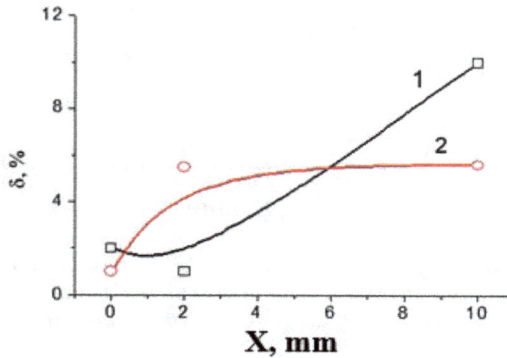

Figure 3.11 *Dependence of the relative content of grains of a ferrite-carbide mixture with globular particles on the distance from the rolling surface. (1) analysis along the central axis; (2) along the fillet [1-3, 123].*

Structure and Properties of Lengthy Rails after Extreme Long-Term Operation
Materials Research Foundations **106** (2021)

Materials Research Forum LLC
https://doi.org/10.21741/9781644901472

Table 3.5 *The relative content of globules in grains of a ferrite-carbide mixture of rails.*

Distance from the surface, mm	ΔV, % (central axis)	ΔV, % (fillet)
0	2	1
2	1	5.5
10	10	5.6

The grains of the ferrite-carbide mixture, which are characterized by a high level of misorientation of the elements of the intragranular structure (fragments), stand apart in this classification (Fig. 3.12, a), as indicated by the quasi-ring character of the microelectron diffraction patterns (Fig. 3.12, b).

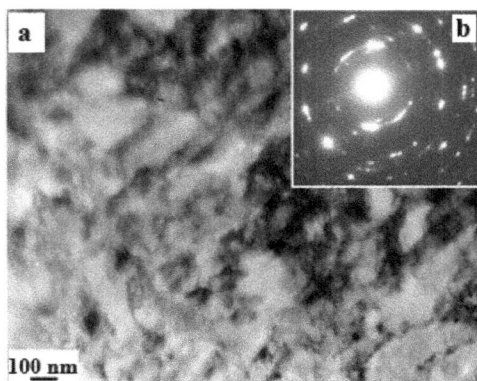

Figure 3.12 *TEM image of grains of a ferrite-carbide mixture (a) bright-field images; (b) micro-electronogram [1-3, 123].*

The azimuthal component of the angle of complete misorientation $\Delta\alpha$ of the substructure elements was estimated according to the method described in [1-3, 123]. The performed estimates showed that $\Delta\alpha \approx 13$ deg. The sizes of fragments of such grains vary from 50 to 100 nm, the sizes of the particles of the carbide phase, having a round shape, from 10 to 30 nm (Fig. 3.12, a). The relative content of these grains is less than one percent.

The ferritic component of the steel structure is defective. The methods of electron microscopy showed a dislocation substructure in the form of chaotically distributed dislocations (Fig. 3.13, a), mesh (Fig. 3.13, b), cellular (Fig. 3.13, c) and fragmented (Fig.

3.13, d) dislocation substructures. In ferrite of pearlite grains, only the first two types of dislocation substructure are observed (substructure of dislocation chaos and latticed dislocation substructure); cellular and fragmented dislocation substructures are revealed only in grains of structurally free ferrite and in grains of a ferrite-carbide mixture [1-3, 123-140].

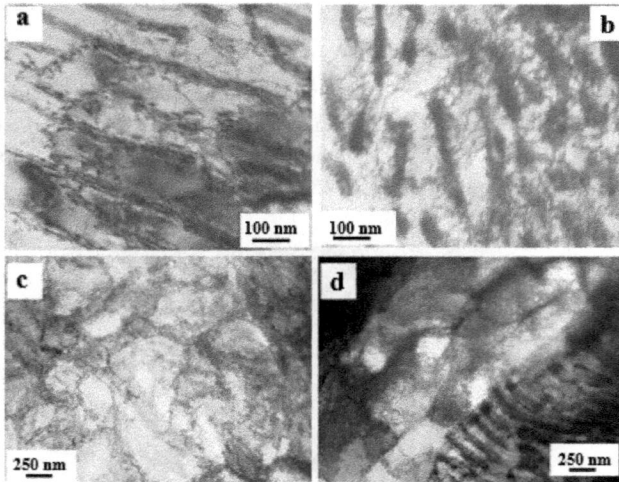

Figure 3.13 *TEM images of the dislocation substructure of the rails; (a) chaotically distributed dislocations; (b) latticed dislocation substructure; (c) cellular dislocation substructure; (d) fragmented dislocation substructure [1-3, 123].*

The scalar dislocation density in the ferrite component of the structure of the investigated steel samples ($<\rho>$) varies in the range from 4×10^{10} to 5.7×10^{10} cm^{-2} (Table 3.6).

Table 3.6 *Scalar dislocation density in the structural components of rails.*

Distance from the surface, mm	Central axis		Fillet	
	ρ (1), 10^{10}, cm^{-2}	ρ (2), 10^{10}, cm^{-2}	ρ (1), 10^{10}, cm^{-2}	ρ (2), 10^{10}, cm^{-2}
0	4.4	5.4	4.3	5.7
2	4.0	5.5	4.3	5.3
10	4.5	4.9	4.2	4.8

Note: ρ (1) is scalar dislocation density in the ferrite component of pearlite grains; ρ (2) is scalar dislocation density in grains of the ferrite-carbide mixture.

The scalar dislocation density reaches its maximum value near the globular particles of the carbide phase and is approximately 5.7×10^{10} cm^{-2}.

Cementite plates of pearlite grains are also defective. An electron microscope image of the substructure of cementite plates of a pearlite colony is shown in Fig. 3.14. Dark-field analysis methods (the image of the structure of the cementite plate was obtained in the reflection of the carbide phase, Fig. 3.14, (b) established that the cementite plates are broken into fragments with sizes 20 - 30 nm. The sizes of the fragments are practically independent of the sample number and the distance of the studied layer from the sample surface.

Electron microscopic studies of the steel structure by the methods of thin foils in transmission made it possible to reveal the bending extinction contours [1-3, 123]. The presence of bending extinction contours in the structure of the material indicates bending-torsion of the crystal lattice of this region of the material and, consequently, internal stress fields that bend the thin foil and, accordingly, strengthen the material.

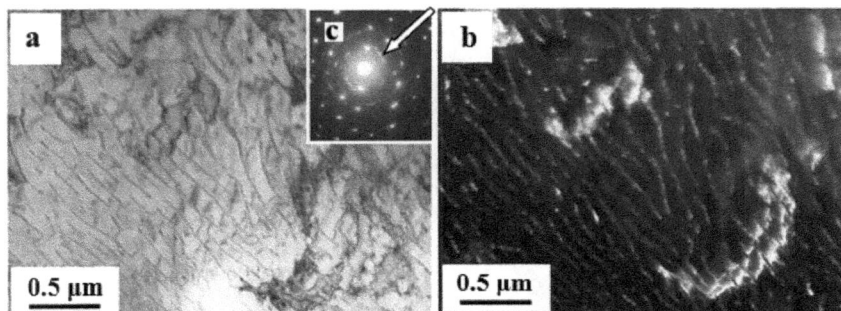

Figure 3.14 *TEM image of cementite plates; (a) bright-field image; (b) dark-field obtained in the reflection [031] Fe₃C + [110] α-Fe; (c) micro-electronogram (the arrow indicates the reflex in which the dark field was obtained) [1-3, 123].*

By analyzing bending extinction contours, one can indicate the sources of internal stress fields and their relative magnitude, i.e. identify stress concentrators. As a result of the studies carried out, it was found that the sources of internal stress fields are the interfaces of pearlite grains (Fig. 3.15, a, b), pearlite grains, and ferrite grains (Fig. 3.15, c). In this case, the contour starts from the grain boundary. Quite a few sources of stress fields are particles of the second phase located along the boundaries and in the volume of grains (Fig. 3.15, d) [1-3, 123-140].

It is generally accepted that the hardening of a material due to bending-torsion of the crystal lattice, caused by internal stress fields, is inversely proportional to the width of the bending extinction contour [1-3, 123]. Consequently, by evaluating the width of the contour, it is possible to estimate the relative value of the hardening of the material introduced by the internal stress fields. The performed estimates show that the average width of the bending extinction contours detected in pearlite grains is 120 nm. The average width of the bending extinction contours detected in grains of structurally free ferrite is 80 nm; in grains of a ferrite-carbide mixture, 75 nm. The minimum width of the bending extinction contours is found near the particles of the carbide phase present in the grains of the ferrite-carbide mixture and is 40 - 50 nm. Consequently, the maximum values of the internal stress fields will be reached near the particles of the second phase. The latter means that the particles of the carbide phase are potential places for the formation of microcracks (they are stress concentrators) and can be dangerous during the operation of rails.

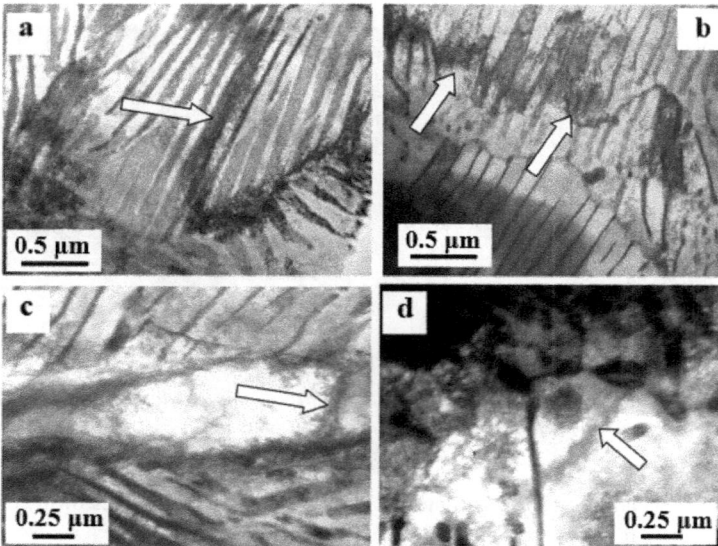

Figure 3.15 *TEM images of the rail structure. Extinction contours are indicated by arrows [1-3, 123].*

Structure and Properties of Lengthy Rails after Extreme Long-Term Operation
Materials Research Foundations **106** (2021)

Materials Research Forum LLC
https://doi.org/10.21741/9781644901472

Conclusion

The performed layer-by-layer studies of the structure, phase composition, and defective substructure of R350HT rails subjected to differential hardening showed that, regardless of the distance to the rolling surface and the direction of analysis (along the central axis or the fillet), hardening is accompanied by the formation of a morphologically multifaceted structure represented by grains of lamellar pearlite, grains of a ferrite-carbide mixture and grains of structurally free ferrite, located in the form of inclusions along the grain boundaries of pearlite. A quantitative analysis has been carried out and the main parameters have been identified that characterize the state of the steel structure [1-3, 123-140]. It is shown that the forming structure has a pronounced gradient character: the state of the surface layer (layer of about 10 mm thick) of the studied rail steel depends on the direction of analysis (along the central axis or on the fillet) and the depth (surface, 2 mm, 10 mm from the surface) of the studied layer. The fact that the relative content of grains of structurally free ferrite and grains of a ferrite-carbide mixture decreases with increasing distance from the cooling surface is common. Consequently, the surface layer of the studied samples of rails is characterized by a relatively more non-equilibrium state of the structure, which is obviously due to the increased rate of its cooling.

4. Evolution of the Structure and Properties of the Metal of Differentially Hardened Rails in the Process of Operation

4.1 Assessment of the quality of differentially hardened rails

4.1.1 Chemical composition of rail steel

Analysis of the results of the verification analysis of the chemical composition of the rail metal after the operation shows that in terms of the content of all chemical elements, the rail metal meets the requirements of GOST R 51685-2013 for E76HF steel.

4.1.2 Mechanical properties of rail steel

Table 4.1 Results of mechanical tests of E76HF rail steel metal.

Sample number	Stress		Elongation, δ	Contraction, ψ	Impact strength, KCU, at a temperature	
	Yield, σ_Y	Ultimate σ_u			+20 ^{0}C	- 60^{0}C*
	N/mm^2		%		J/cm^2	
1	830	1260	11.5	39	30	15*
2	820	1270	11.5	40	34	16*
Requirements of GOST R 51685-2013 for rails of category R350HT (* R350HTNN)	Not less					
	800	1180	9.0	25.0	15.0	15.0*

As can be seen from the test results presented in Table 4.1, in terms of the level of mechanical properties in tension, the metal of the rails after the operation meets the requirements of the standard for rails of the R350HT and R350HTNN steel categories.

Analysis of the results (Table 4.2) shows that the hardness of the rail after the operation meets the requirements of GOST R 51685-2013 for rails of the R350HT category.

Table 4.2 Results of determining the hardness of R350HT rails.

Material	Hardness, HB							
	On the rolling surface of the head	On the cross-section of the head				In the neck	In the foot	
		10 mm			22 mm		1	2
		Fillet left	Center	Fillet right				
Rail R350HT	395	363	375	363	354	331	339	343
Requirements of GOST R 51685-2013 for rails of category R350HT	352-405	Not less					Not more	
		341			321	341	363	

4.1.3 Macrostructure of rail steel

The macrostructure of the metal was analyzed following the requirements of GOST R 51685-2013 on a full-profile template (Fig. 4.1). The analysis presented in Fig. 4.1 images of rail steel allows us to conclude that the macrostructure of the rail metal after the operation in terms of axial segregation, point inhomogeneity, and segregation strips is satisfactory. Macrostructure defects unacceptable according to GOST R 51685-2013 (flocs, exfoliation, cracks, crusts, spotted liquation, foreign metal, and slag inclusions) were not found. From the rolling surface of the head, after etching, a network of small cracks of contact-fatigue origin was discovered (Fig. 4.1, b). The surface of the rail does not have rolled out dirt, cracks, flaws, rolled crusts, scabs, crimps, scale shells, undercuts, dents, burrs, transverse marks, and transverse scratches, not allowed by GOST R 51685-2013.

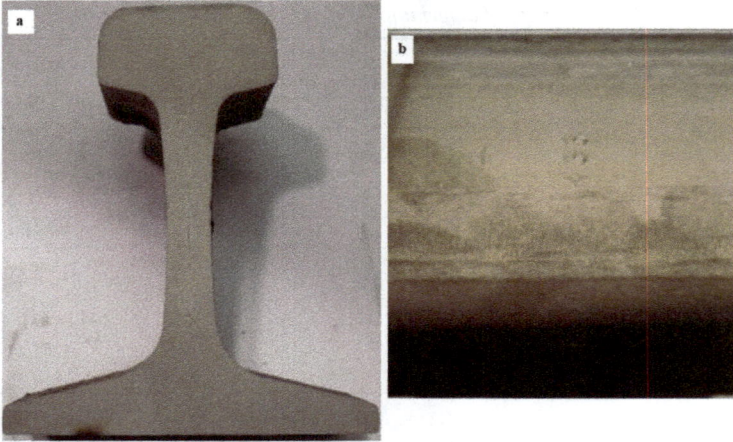

Figure 4.1 Macrostructure of metal of differentially heat-strengthened rails after the operation [2, 123].

On the side of the working fillet, the rolling surface of the head is more worn out, on the side face where there is a melting of metal (Fig. 4.2).

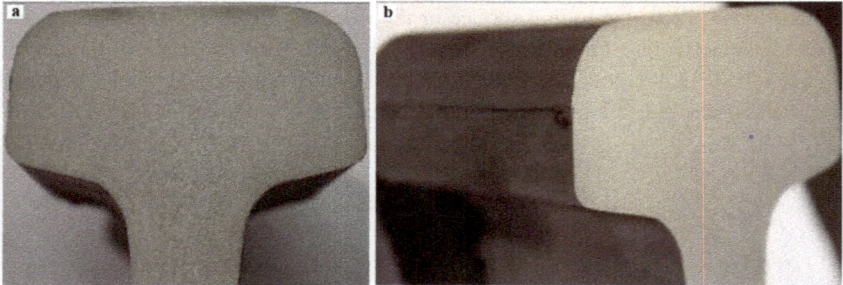

Figure 4.2 Macrostructure of the metal of the working fillet of differentially heat-strengthened rails after operation [2, 123].

Optionally, the rail metal was assessed following GOST 1778-70. The metal contains sulfides, point oxides, and brittle silicates up to 1.0 points. Nitrides, carbonitrides, plastic and non-deformable silicates, string oxides were not found in the metal under study [2, 123].

4.1.5 Microstructure of rail steel after operation [2, 123]

On non-etched thin sections at the place of defects of contact-fatigue origin, discontinuities are observed, filled with corrosion products, passing at an acute angle to the surface to a depth of 140 μm (Fig. 4.3). The distance between defects varies from 700 μm to 1100 μm (Fig. 4.4).

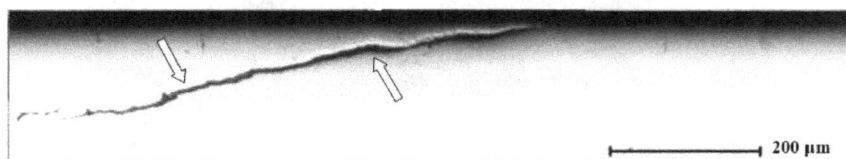

Figure 4.3 Microstructure of non-etched sections of rail steel after the operation. Arrows indicate a surface defect in the form of an extended crack [2, 123].

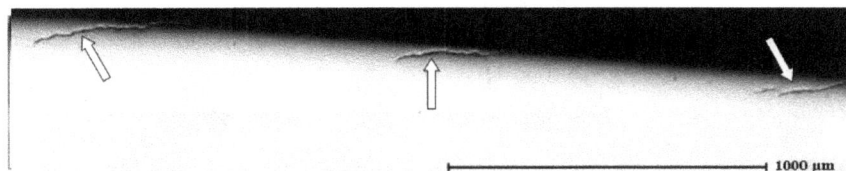

Figure 4.4 Surface defects in the form of extended cracks (indicated by arrows), found in the study of the microstructure of non-etched sections of rail steel after the operation [2, 123].

On etched thin sections from the surface of the working fillet, a significantly deformed structure is observed to a depth of 200 μm. The size of the decarburized surface layer, identified by a continuous ferrite grid, does not exceed 250 μm (Fig. 4.5).

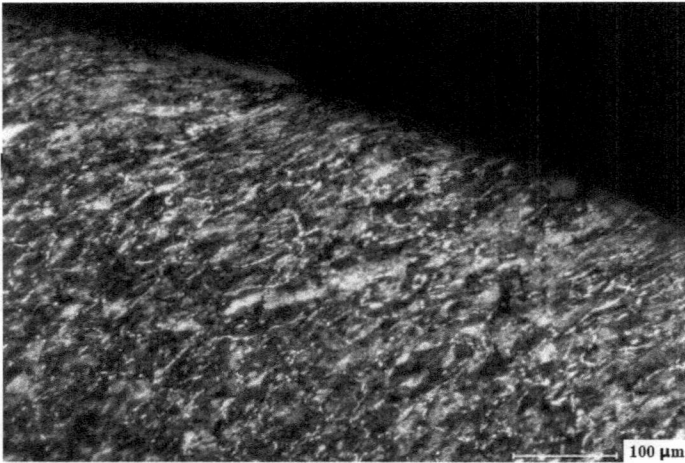

Figure 4.5: *Microstructure of the etched thin sections of the surface of the rail steel fillet after the operation [2, 123].*

Local light-etched areas of work-hardened metal are observed from the rolling surface, along which cracking develops (Fig. 4.6).

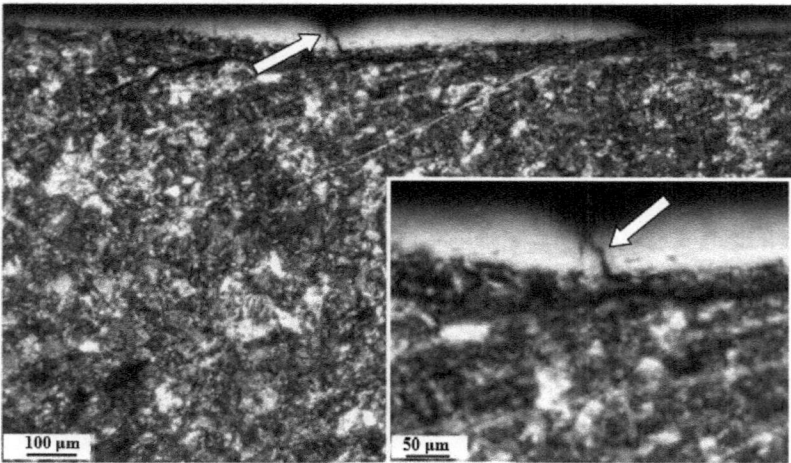

Figure 4.6 *Microstructure of etched sections of the rolling surface of rail steel after the operation; the arrow indicates a microcrack [2, 123].*

The microstructure of the rail metal after the operation is satisfactory and is lamellar pearlite of 2-3 points of the GOST 8233 scale one with scattered ferrite areas along the grain boundaries. The number of ferrite grains is less than 5% and is evaluated by a point of 1.5 on the scale seven of GOST 8233. Bainite was not detected in the microstructure of the rail under study. With the distance from the surface of the head, the pearlite rail acquires a coarser structure, which indicates a decrease in the cooling rate of the material with distance from the processing surface. Typical images of the structure of an etched transverse section of rail steel after the operation at various distances from the rail rolling surface are shown in Figs. 4.7 - 4.9.

Figure 4.7 *Microstructure formed after the operation of rails at a distance of 2 mm from the rolling surface [2, 123].*

Thus, according to the results of qualitative assessment of differentially hardened rails of the R350HT category done according to the methods [2, 123], the following observations were made:

1. In terms of the content of chemical elements, the metal meets the requirements of GOST R 51685-2013 for steel grade E76HF;

2. In terms of the level of mechanical properties, hardness, impact strength, contamination with non-metallic inclusions, macro- and microstructure, the quality of the metal meets the requirements of GOST R 51685-2013 for rails of the R350HT and R350HTNN categories;

3. The defects of contact-fatigue origin formed on the rolling surface of the head during the operation have an insignificant degree of development, their depth does not exceed 140 microns. It is recommended to carry out preventive grinding of rails to eliminate operational defects of contact-fatigue origin.

The grain size of the studied metal is estimated to be seven on the scale two of GOST 5639-82 (Fig. 4.10).

Figure 4.8 *Microstructure formed after the operation of rails at a distance of 10 mm from the rolling surface [2, 123].*

Materials Research Forum LLC
https://doi.org/10.21741/9781644901472

Figure 4.9 *Microstructure formed after the operation of rails at a distance of 22 mm from the rolling surface [2, 123].*

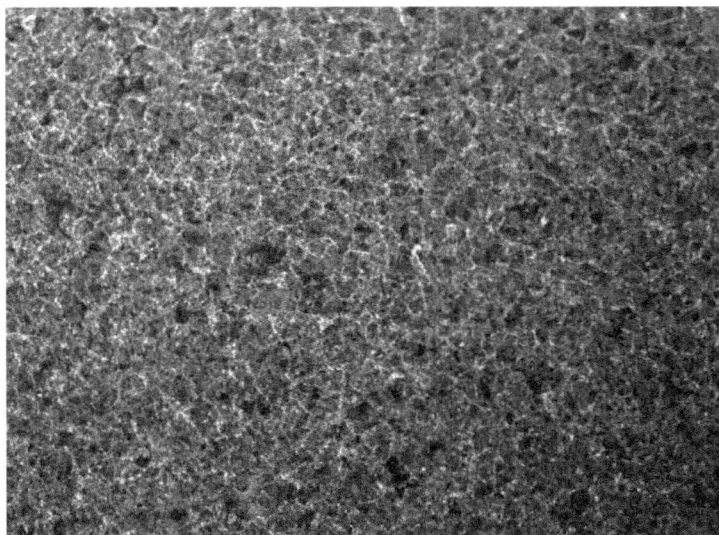

Figure 4.10 *Polycrystalline (grain) structure observed after the operation of rails (magnification 100 times) [2, 123].*

4.2   Phase composition and defective substructure of differentially hardened rails after operation [141-146]

The main structural component of the investigated steel, as revealed by the methods of optical metallography of the etched section (Figs. 4.7 - 4.9), is pearlite of plate morphology, a characteristic image of which is shown in Fig. 4.11. This image was obtained by transmission electron diffraction microscopy methods of a foil located at a distance of 10 mm from the rolling surface.

According to the classical definition, pearlite is a structural component of steels and cast irons, i.e. a eutectoid mixture of ferrite and cementite [2, 123]. During crystallization under normal conditions, pearlite has a lamellar structure consisting of alternating plates of pearlite and cementite [145, 146].

The studies carried out in this work demonstrated numerous imperfections in the structure of lamellar pearlite. Namely, there is an alternating structure ("comb" type structure, indicated by arrows in Fig. 4.12, a) and breaks of cementite plates (ferrite bridges) (Fig. 4.12, b).

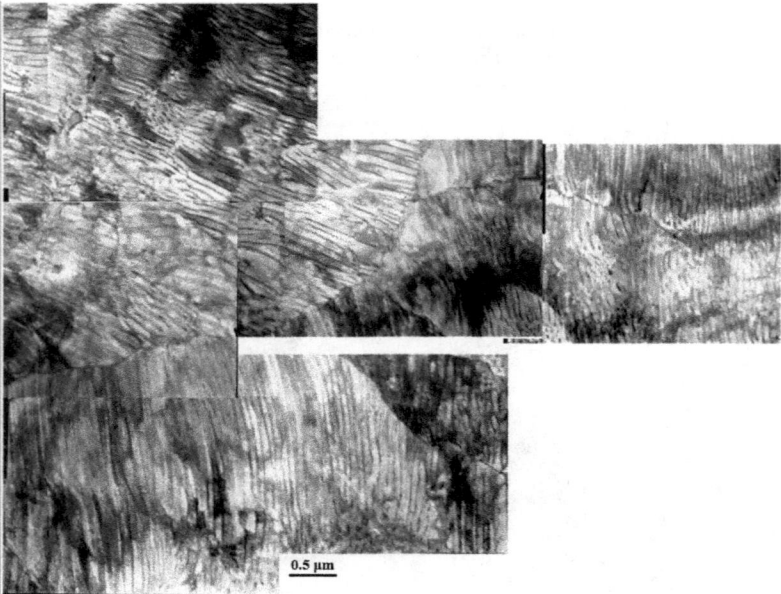

Figure 4.11 Electron microscopic image of the structure of lamellar pearlite at a distance of 10 mm from the rolling surface along the central axis [141-146].

62

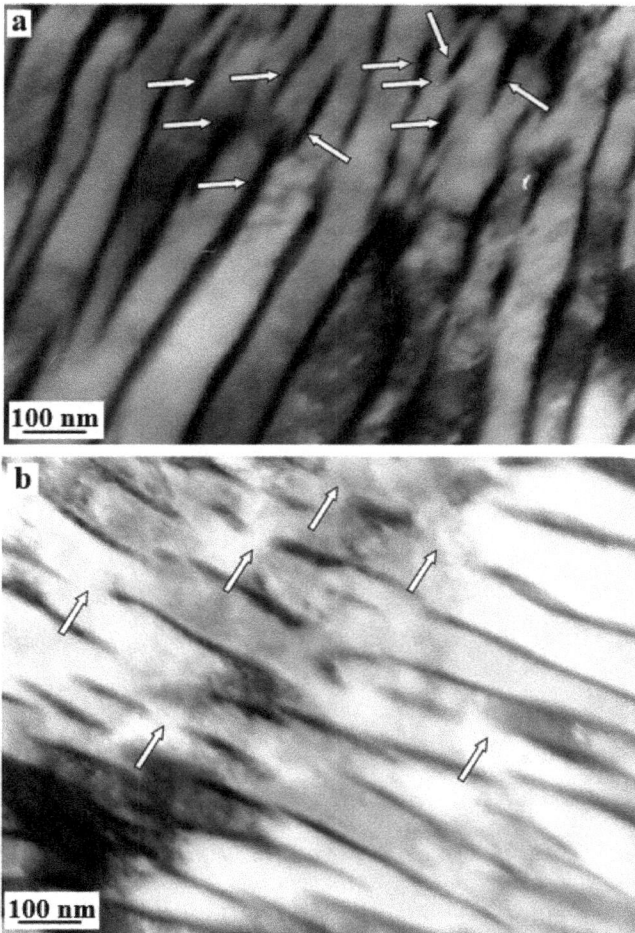

Figure 4.12 *Electron microscopic image of the lamellar pearlite structure; (a) "comb"*
type structure; (b) ferrite bridges (indicated by arrows) [141-146].

Quite often, curved cementite plates (Fig. 4.13, a) and cementite plates of variable
thickness (Fig. 4.13, b) are observed in a pearlite colony.

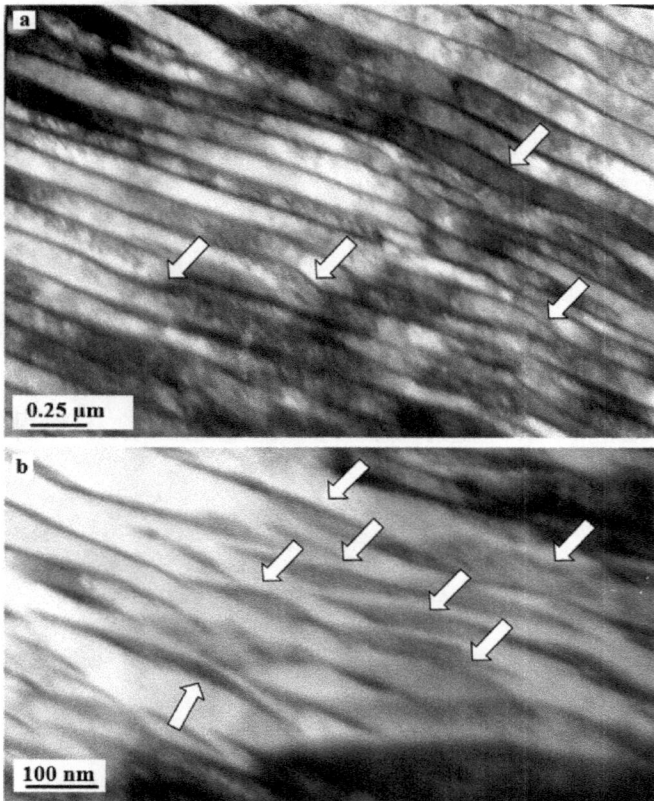

Figure 4.13 TEM image of the structure of lamellar pearlite of rail steel after the operation; (a) curved cementite plates (indicated by arrows); (b) cementite plates of variable thickness (indicated by arrows) [141-146].

In some cases, aggregates of cementite plates of various configurations are found in the structure of pearlite colonies (Fig. 4.14).

Structure and Properties of Lengthy Rails after Extreme Long-Term Operation
Materials Research Foundations **106** (2021)

Materials Research Forum LLC
https://doi.org/10.21741/9781644901472

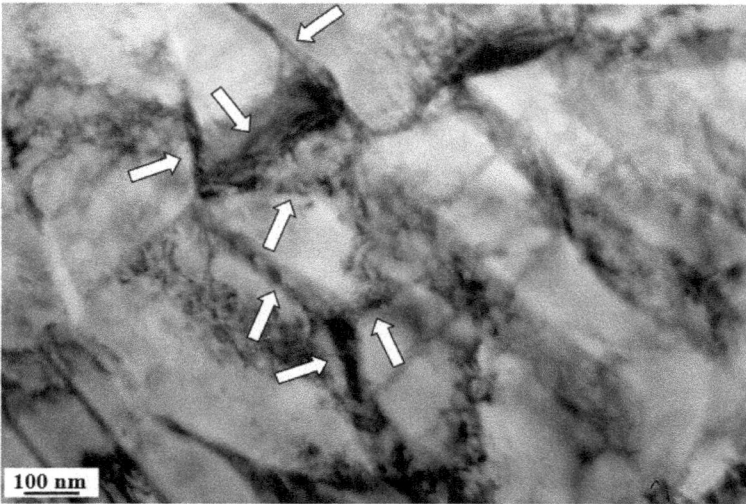

Figure 4.14 *Electron microscopic image of pearlite grains of lamellar morphology, with intergrown cementite plates (indicated by arrows) [141-146].*

Quite often, ferrite plates of pearlite colonies have alternating light and gray contrast. This diffraction contrast indicates that the ferrite plates are divided, apparently as a result of elastic stresses, into weakly misoriented regions (Fig. 4.15). The azimuthal component of the angle of complete misorientation of such areas, discovered in the analysis of microelectron diffraction patterns (the method for determining misorientation is described in detail in [2, 123]), is 1–2 degrees. The most distinct areas of misorientation of ferrite plates of a pearlite colony are obtained by the methods of dark-field analysis, the results of which are shown in Fig. 4.16. The regions of misorientation have an elongated shape, the longitudinal dimensions of the regions vary from 200 to 800 nm, and the transverse dimensions of the regions are in most cases equal to the transverse dimensions of ferrite plates and are on average 250 nm.

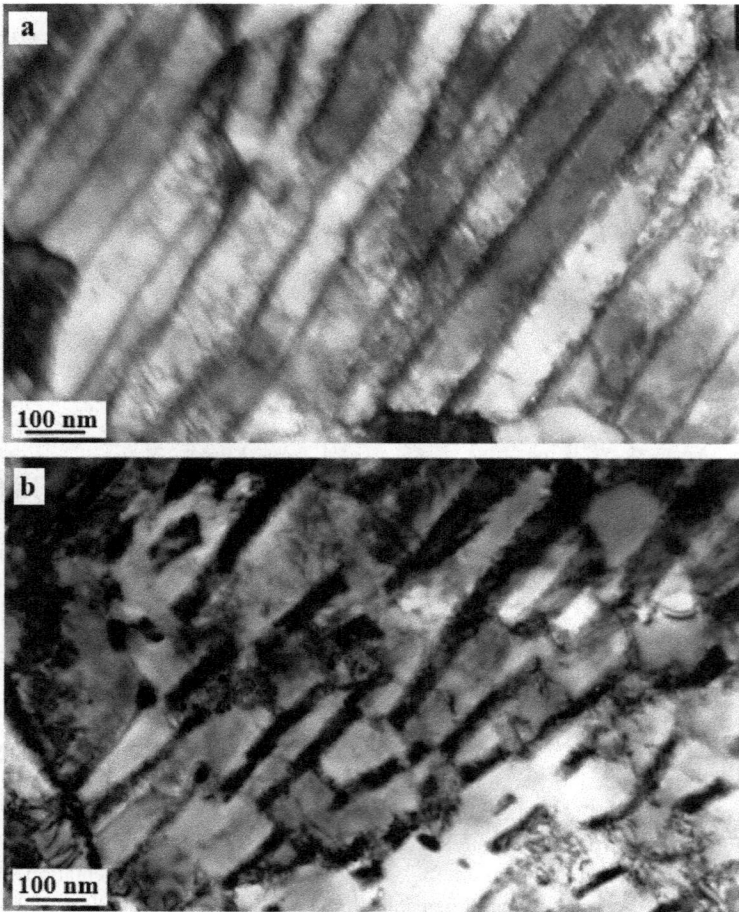

Figure 4.15 *Diffraction contrast on ferrite plates [141-146].*

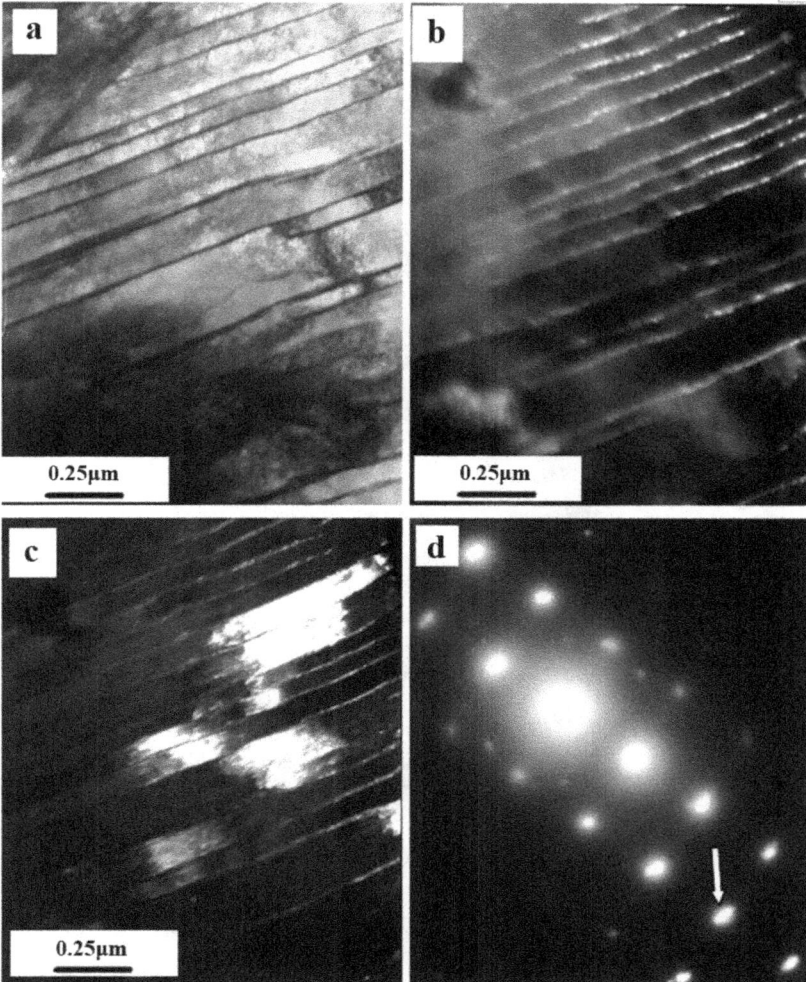

Figure 4.16 *Electron microscopic image of areas of misorientation of ferrite plates; (a)
bright field; (b, c) dark fields obtained in the reflections [012] Fe₃C and [102] Fe₃C +
[110] α-Fe; (d) micro-electronograms (the figures indicate the reflections in which dark
fields were obtained: one for (b), two for (c); the arrow indicates a reflex of the {420} α-
Fe type, the azimuthal broadening of which indicates the disorientation of ferrite regions)
[141- 146].*

In the volume of ferrite plates of pearlite colonies, a dislocation substructure is revealed, a characteristic electron microscopic image of which is shown in Fig. 4.17.

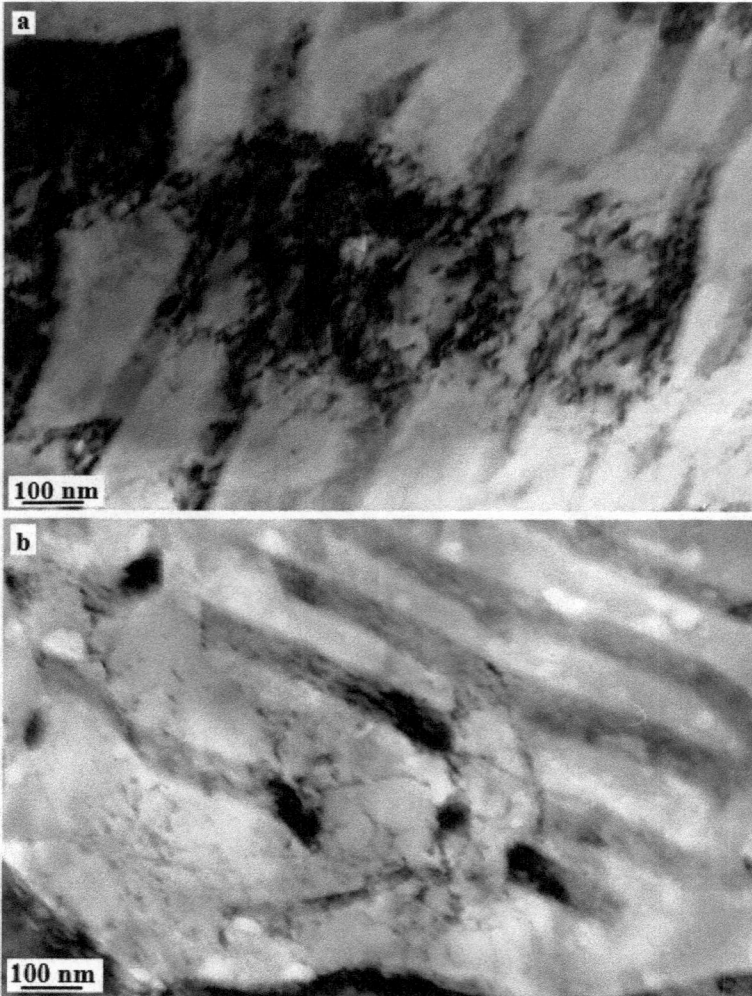

Figure 4.17 *Electron microscopic image of the dislocation substructure formed in ferrite pearlite plates [141-146].*

Structure and Properties of Lengthy Rails after Extreme Long-Term Operation
Materials Research Foundations **106** (2021)

Materials Research Forum LLC
https://doi.org/10.21741/9781644901472

According to the morphological feature in the studied steel [112, 118, 147, 148], a mesh dislocation substructure is revealed (Fig. 4.17, a) as well as the substructure of dislocation chaos (Fig. 4.17, b).

Analysis of the steel structure by transmission electron microscopy of thin foils demonstrated the presence of bending extinction contours in the electron microscopic images of pearlite grains [2, 123], a typical image of which is shown in Fig. 4.18.

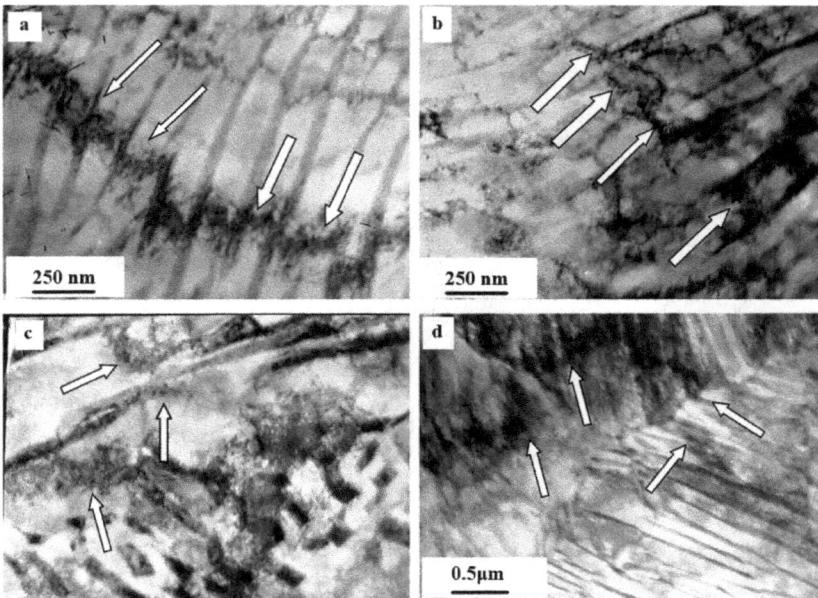

Figure 4.18 *Electron microscopic image of bending extinction contours (indicated by arrows) [141-146].*

The presence of bending extinction contours indicates bending-twisting of the crystal lattice of the analyzed foil region [2, 123]. The performed studies show that the sources of bending-torsion of the crystal lattice (stress concentrators) are mainly the interfaces between ferrite and cementite plates. In most of the observed cases, the contours are located perpendicular to the interface (Fig. 4.18, a). The source of the curvature-torsion of the crystal lattice of the material can also be the ends of the cementite plates (Fig. 4.18, b, c), as well as the interfaces of pearlite grains (Fig. 4.18, d).

It was shown by metallography of etched sections (Fig. 4.10) that grains of structurally free ferrite are present in the steel structure, i.e. ferrite grains, in the volume of which there are no particles of the carbide phase. The relative content of such grains is small and does not exceed 5% of the steel structure. Ferrite grains, as a rule, are arranged in the form of interlayers along the grain boundaries of pearlite (Fig. 4.19) or at the joints of the grain boundaries of pearlite (Fig. 4.20).

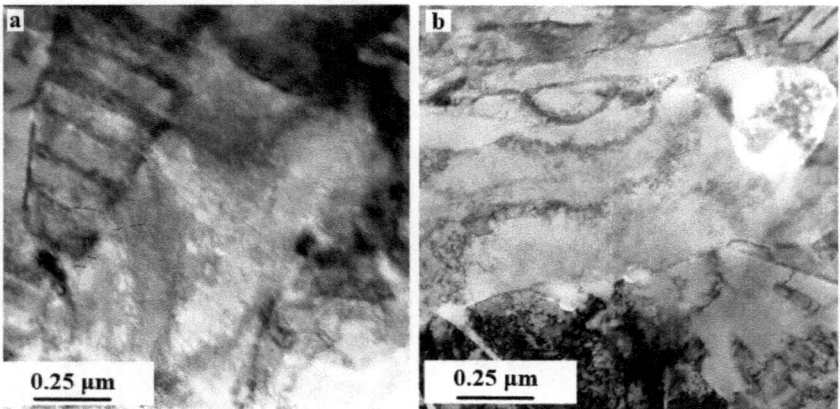

Figure 4.19 *TEM image of grains of structurally free ferrite [141-146].*

In the volume of grains of structurally free ferrite, as well as in grains of lamellar pearlite, bending extinction contours are present in electron microscopic images (Figs. 4.19 - 4.21). The latter, as noted above, indicates the curvature-torsion of the crystal lattice of ferrite. As a rule, bending contours begin at the grain boundary. The latter indicates that the sources (concentrators) of stresses are located precisely at the boundaries of ferrite grains. It should be noted that bending contours are usually formed simultaneously from the boundary of two adjacent grains, i.e. both contacting grains are in an elastic-plastic state (in our case (Fig. 4.21) pearlite grain and ferrite grain).

In the volume of grains of structurally free ferrite, as well as in the ferrite interlayers of pearlite colonies, a dislocation substructure is observed (Fig. 4.22). In most of the cases revealed, the dislocations are randomly distributed over the volume of the ferrite grain; in rare cases, a reticular dislocation substructure is observed. As a rule, the dislocation density depends on the distance to the grain boundary: in the bulk of the grain, the dislocation density is lower than at the boundaries. On average over the volume of a ferrite grain, the value of the scalar dislocation density varies within $(1.5 - 3.5) \times 10^{10}$ cm^{-2}.

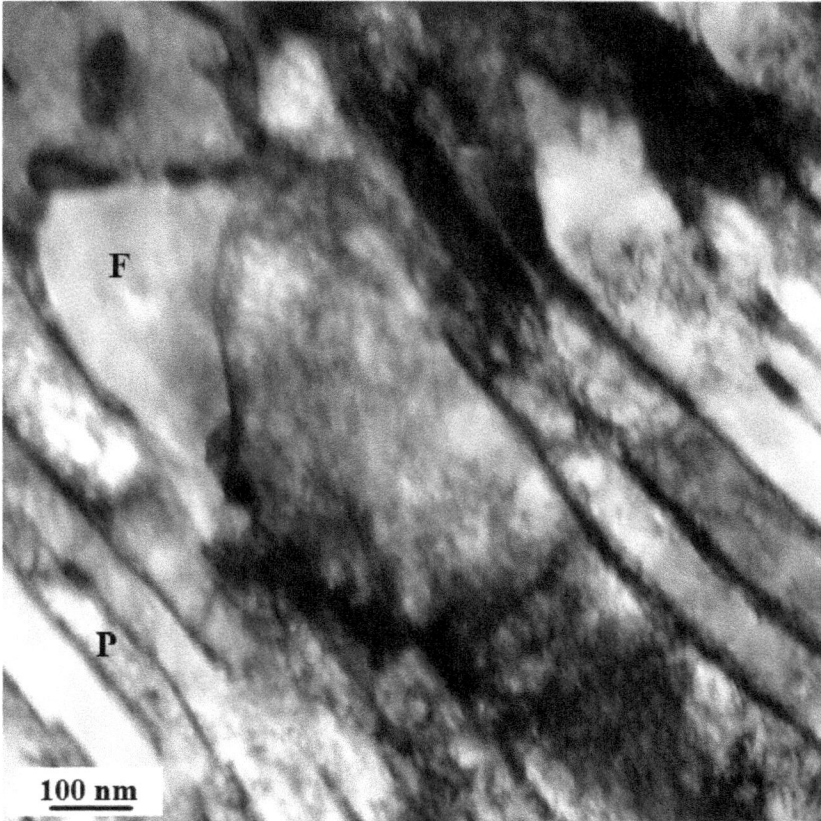

Figure 4.20 *TEM image of a grain of structurally free ferrite (indicated in the figure by "F"), located at the junction of pearlite grains ("P") [141-146].*

Along with the grains of lamellar pearlite, the studied rail steel contains ferrite grains, in the volume of which there are cementite particles of a globular shape, located randomly in the grain. In this work, such structural formations will be referred to as grains of a ferrite-carbide mixture. The relative content of such grains in the steel does not exceed 5%. A typical image of grains of a ferrite-carbide mixture is shown in Fig. 4.23. As a rule, grains of a ferrite-carbide mixture are located in islands between grains of lamellar pearlite (Fig. 4.23, a; the grain is outlined by an oval). The sizes of cementite particles in such grains vary from 30 to 100 nm (Figs. 4.23 (b) and 4.24).

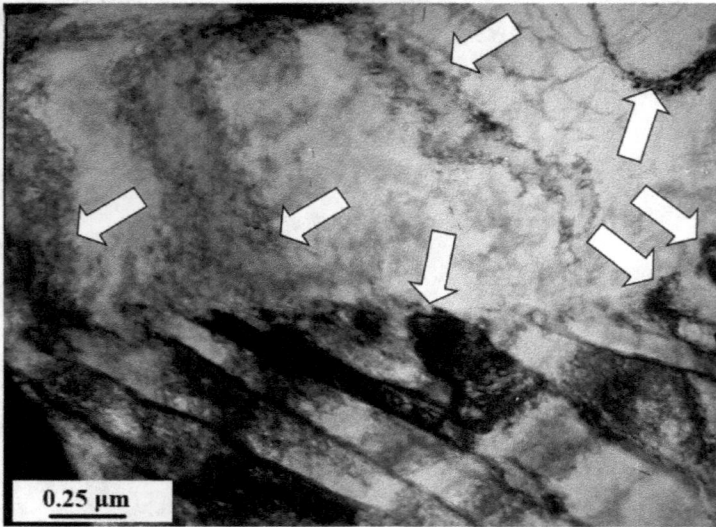

Figure 4.21 *Electron microscopic image of bending extinction contours (indicated by arrows) in a grain of structurally free ferrite [141-146].*

In the volume of grains of the ferrite-carbide mixture, a dislocation substructure is observed, a typical image of which is shown in Fig. 4.25. A network dislocation substructure (Fig. 4.25, a) and a substructure of dislocation chaos (Fig. 4.25, b) are observed. The scalar dislocation density of grains of the ferrite-carbide mixture is 3.2×10^{10} cm^{-2}.

The electron microscopic images of the grains of the ferrite-carbide mixture have bending extinction contours (Figs. 4.23 - 4.25). The contours generally begin and end on particles of the carbide phase. Consequently, the main sources of curvature-torsion of the crystal lattice in such grains are precisely the particles of the second phase.

Figure 4.22 *TEM image of the dislocation substructure of grains of structurally free ferrite [141-146].*

It should be noted that the presented characteristics of grains of a ferrite-carbide mixture were obtained by studying foils located at a distance of 10 mm from the rolling surface along the central axis. As will be shown below, similar structures are formed in the surface layer of the rail metal during long-term operation as a result of the thermal-deformation transformation of lamellar pearlite.

Thus, the results of electron microscopic micro-diffraction studies of the phase composition and defective substructure of the rail metal presented in this section have shown that regardless of the direction of the analysis (rolling surface, central axis roll, or fillet, along the radius of curvature), steel is a morphologically diverse material [141-146]. The main type of structure is perlite lamellar morphology; grains of structurally free ferrite and grains of a ferrite-carbide mixture are present in a substantially smaller amount in steel. Obviously, the long-term operation of rails under conditions of intense impact on the rolling surface will be accompanied by the formation of a gradient structure in the material. The characteristics of this structure are discussed below.

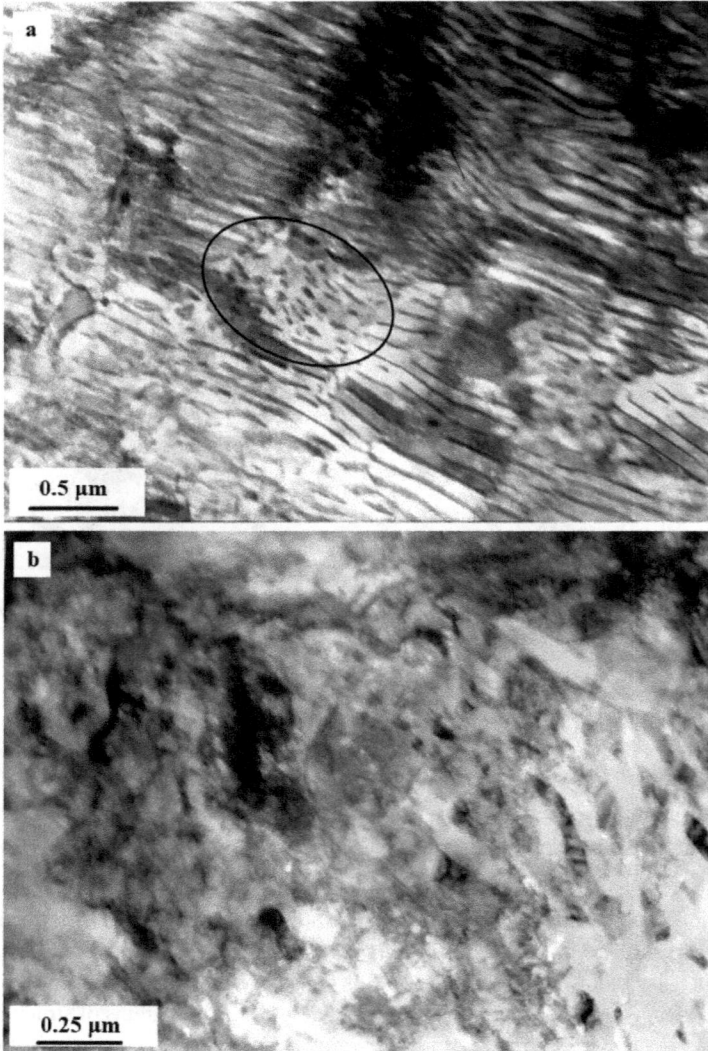

Figure 4.23 *TEM image of a grain of a ferrite-carbide mixture (a; the grain is outlined with an oval) and its defective substructure (b) [141-146].*

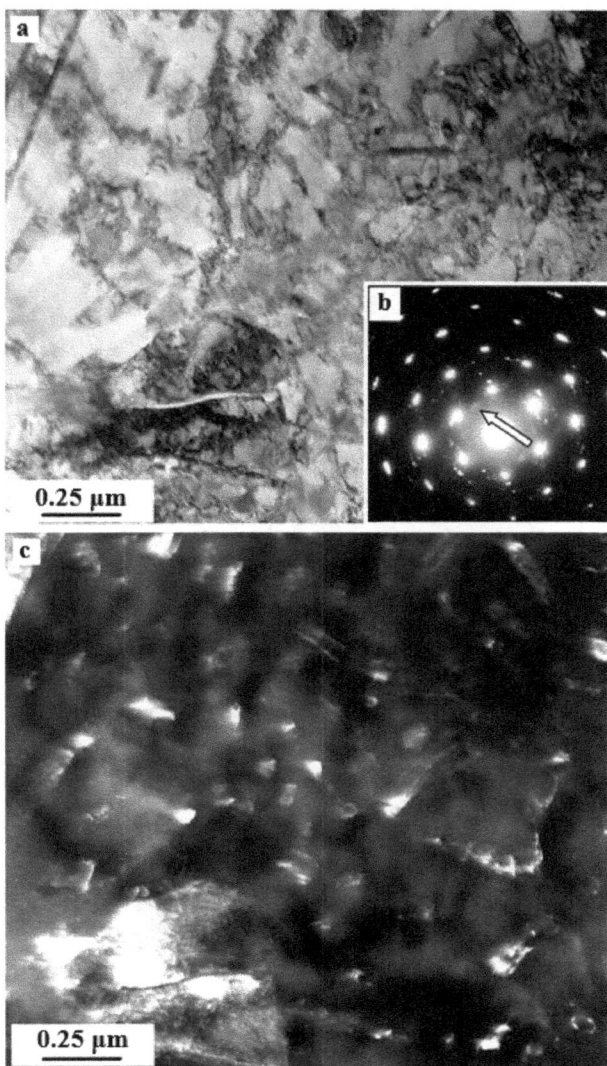

Figure 4.24 *TEM image of a grain of a ferrite-carbide mixture; a - bright field; b –*
micro-electronogram; (c) dark field obtained in the [201] F3C reflex (the reflex is
indicated by an arrow in the micro-electronogram) [141-146].

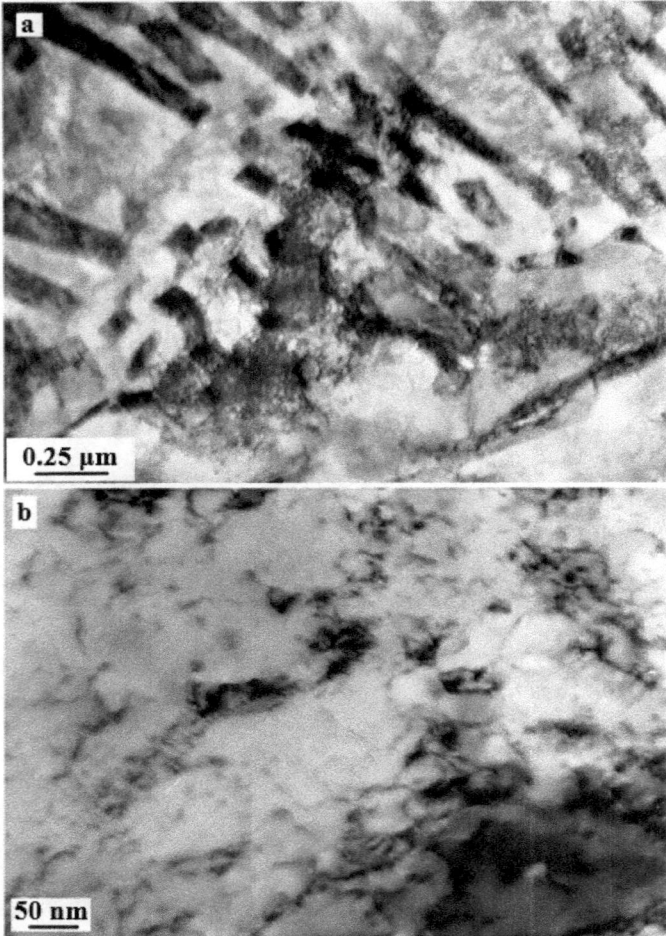

Figure 4.25 *TEM image of the dislocation substructure of grains of a ferrite-carbide mixture [141-146].*

4.3 Gradients of the structural-phase state of steel, formed as a result of the operation of rails [150]

A quantitative analysis of layer-by-layer electron microscopic studies of the phase composition and defective substructure of steel made it possible to consider the issue of the formation of structural-phase state gradients in rails during operation.

Long-term operation of rails is almost always accompanied by deformation transformation of the material structure [46-49, 121, 149]. The quantitative analysis of the morphological state of the steel structure carried out in this work showed that the operation of rails is accompanied by the transformation of predominantly the state of the grains of lamellar pearlite. Namely, the destruction of cementite plates. Fig. 4.26 shows the dependence of the relative content of destroyed and not destroyed pearlite on the distance to the rail surface.

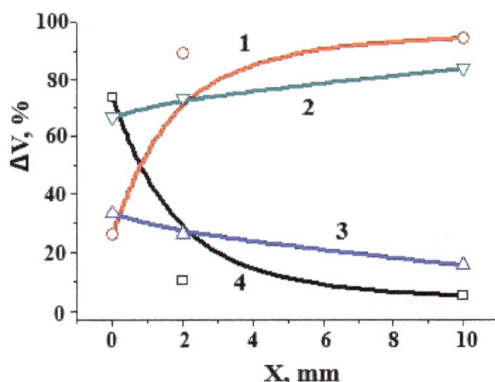

Figure 4.26 *Dependence of the relative content of pearlite not destroyed (curves 1 and 2) and destroyed (curves 3 and 4) on the distance from the contact surface of the rails. Curves 1 and 4 represent rolling surface (central axis); (2, 3) represent working fillet [2, 123, 150].*

Typical examples of the resulting structure are shown in Fig. 4.27.

Analyzing the results presented in Fig. 4.26, it can be noted that, regardless of the position of the analyzed volume (rolling surface or fillet), the destruction of the lamellar pearlite structure is maximum in the surface layer of the rail with a thickness of no more than 2 mm. However, the degree of destruction of the lamellar pearlite structure essentially depends on the position of the analyzed volume; namely, the relative content of grains of destroyed pearlite is more than two times higher on the rolling surface than in the surface layer of the working fillet [2, 123, 150].

The operation of the rails is accompanied, as noted above, by the destruction of pearlite grains. One of the main mechanisms of destruction of pearlite grains during the plastic deformation of steel is the cutting of cementite plates by sliding dislocations [121, 150]. Fig. 4.28 shows the gradient of the scalar dislocation density in fractured and non-fractured grains of lamellar pearlite.

It can be seen that, with distance from the contact surface, the structure of the working fillet is characterized by a decrease in the scalar dislocation density regardless of whether the structure of pearlite colonies is destroyed or not (Fig. 4.28, curves 2 and 3). The gradient of the scalar dislocation density along the central axis of the rail has a different character. Namely, in pearlite grains destroyed during the operation of rails, the scalar dislocation density is maximum on the contact surface; in pearlite grains not destroyed during the operation of rails, the maximum value of the scalar dislocation density is observed in the volume of the rails (at a distance of 10 mm from the contact surface).

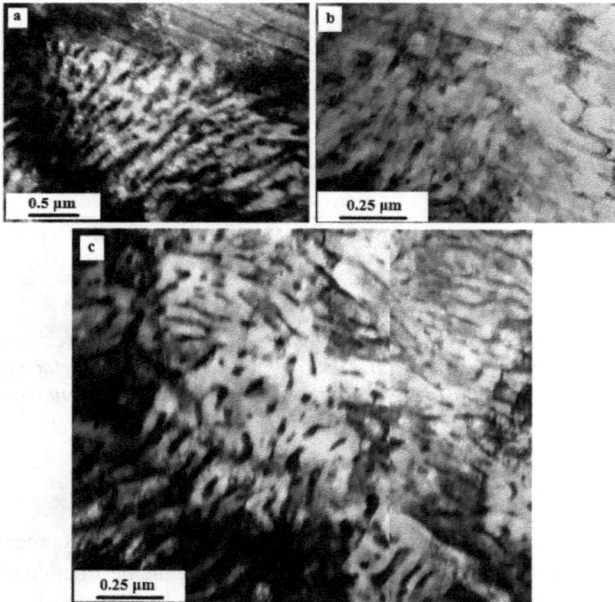

Figure 4.27 *Electron microscopic image of the structure of destroyed lamellar pearlite of rail steel [2, 123, 150].*

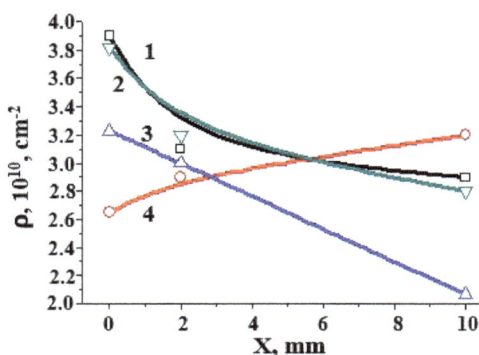

Figure 4.28 *Dependence of the scalar dislocation density of the ferrite component of pearlite colonies, in destroyed (curves 1 and 3) and not destroyed (curves 2 and 4) states, on the distance from the rolling surface. Curves 1 and 4 represent the structure of the rolling surface along the central axis; curves 2 and 3 - the structure along the working fillet [2, 123, 150].*

The operation of rails is accompanied by an increase in the level of elastic-plastic stresses in steel. The magnitude of the elastic-plastic stresses of steel, according to [121, 150-154], will be characterized by the excess density of dislocations and the amplitude of curvature-torsion of the crystal lattice of the material. Noteworthy is that both of these characteristics of steel are determined by analyzing the bending extinction contours of the material (the analysis technique is described in detail in Chapter 2 of this monograph). The gradient of the excess dislocation density in fractured and undisturbed grains of lamellar pearlite is shown in Fig. 4.29.

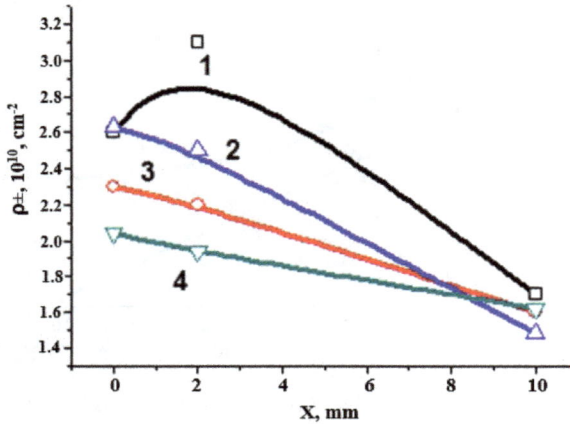

Figure 4.29 *Dependence of the excess dislocation density of the ferrite component of pearlite colonies in destroyed (curves 1 and 2) and not destroyed (curves 3 and 4) states on the distance from the rolling surface. Curves 1 and 3 represent the structure of the rolling surface along the central axis; curves 2 and 4 - structure along the working fillet [2, 123, 150].*

Analyzing the results presented in Fig. 4.29, it can be noted that the maximum value of the excess dislocation density of the ferrite component of pearlite colonies is recorded at the contact surface of the fillet (regardless of the state of the pearlite colonies: destroyed or not) and the rolling surface (for undisturbed pearlite colonies). For the destroyed pearlite colonies of the structure of the working surface of the rails, the maximum value of the excess density of dislocations of the ferrite component is achieved in the layer located at a distance of 2 mm from the contact surface (Fig. 4.29, curve 1) [2, 123, 150].

Fig. 4.30 shows the results of the analysis of the gradient of the amplitude of curvature-torsion of the crystal lattice of steel formed during the operation of rails.

Analyzing the results presented in Fig. 4.30, we can note a similar nature of the dependencies with the results shown in Fig. 4.29. Namely, the maximum value of the amplitude of the curvature-torsion of the crystal lattice of the ferrite component of pearlite colonies is fixed at the contact surface of the fillet (regardless of the state of the pearlite colonies, destroyed or not) and the rolling surface (for undisturbed pearlite colonies). For the destroyed colonies of pearlite of the structure of the working surface of the rail, the maximum value of the amplitude of the curvature-torsion of the crystal lattice

Materials Research Forum LLC
https://doi.org/10.21741/9781644901472

of the ferrite component is achieved in a layer located at a distance of 2 mm from the contact surface (Fig. 4.30, curve 1).

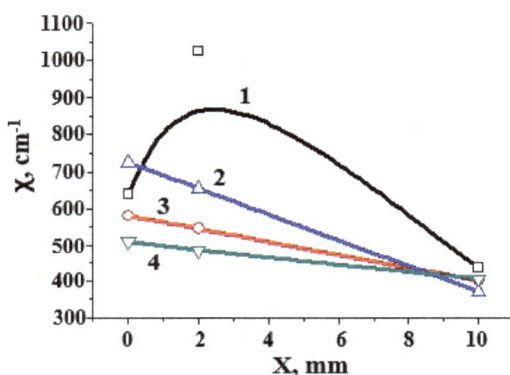

Figure 4.30 *Dependence of the amplitude of the curvature-torsion of the crystal lattice of the ferrite component of pearlite colonies in destroyed (curves 1 and 2) and not destroyed (curves 3 and 4) states on the distance from the rolling surface. Curves 1 and 3 - the structure of the rolling surface along the central axis; curves 2 and 4 — structure along the working fillet [2, 123, 150].*

Thus, the results of a quantitative analysis of the structure parameters indicate the formation of a gradient substructure, which is expressed in a regular change in the scalar and excess dislocation density, the amplitude of curvature-torsion of the crystal lattice of steel, and the degree of deformation transformation of the lamellar pearlite structure. It has been shown that the destruction of the cementite plates of pearlite colonies proceeds mainly by two mechanisms - by cutting and sliding dislocations and as a result of the escape of carbon atoms from the crystal lattice of cementite to dislocations [2, 123, 150].

4.4 Transformation of the structure of lamellar pearlite of steel during long-term operation of rails [155, 156]

The above results of the study of the rail metal after long-term operation indicate the transformation of the structure of lamellar pearlite. In the research literature, two mechanisms of destruction of cementite plates during the deformation of steel with a pearlite structure are mainly discussed.

The first of them consists in cutting the plates by moving dislocations and carrying carbon atoms into ferrite in the field of dislocation stresses. The estimates carried out in [157] show, that in this case, the maximum effect of cementite decomposition cannot exceed tenths of a percent of the available amount of cementite.

The second mechanism consists in the pulling of carbon atoms from the lattice of the carbide phase during the plastic deformation by dislocations. Because of a noticeable difference in the average binding energy of carbon atoms with dislocations (0.6 eV) and with iron atoms in the cementite lattice (0.4 eV) this process leads to the formation of Cottrell atmospheres [121, 150]. Diffusion of carbon occurs in the stress field created by the dislocation substructure that forms around the cementite plate. In this case, the degree of decomposition of cementite should be determined by the value of the dislocation density and the type of substructure. So, according to the authors of works [157, 158], the model of cementite decomposition can be presented as follows. Plastic deformation of pearlitic steel causes the formation of a cellular substructure with cell boundaries located at the cementite-ferrite interface. In the presence of a thermodynamic stimulus (the binding energy of carbon atoms with dislocations is higher than with iron atoms in cementite), carbon atoms, the mobility of which is initiated by plastic deformation, are transferred from the surface layers of cementite to dislocations localized at the interface. The results obtained, according to [157, 158], indicate an unambiguous relationship between the density and the nature of the distribution of dislocations in ferrite, on the one hand, and the degree of decomposition of cementite, on the other, confirming the dislocation nature of decomposition.

Thus, the authors of [155, 156] conclude that the main contribution to the destruction of lamellar cementite is made by the plastic deformation-accelerated transfer of carbon atoms from the cementite lattice to dislocations in ferrite located at the ferrite-cementite interface.

Summarizing, based on the literature data of the studies performed above, it can be stated that the plastic deformation of a pearlitic steel is accompanied by two processes simultaneously: cutting of cementite plates and their dissolution. The first process, which is carried out by the mechanism of cutting carbide particles and pulling apart their fragments, is accompanied only by a change in their linear dimensions and morphology (Fig. 4.31 and Fig. 4.32).

The change in the elemental composition of cementite during crushing is minimal. During the second process (the action of the dissolution mechanism "in place"), a completely different picture is observed. At the initial stage of transformation, the cementite plates of the pearlite colony are entangled with sliding dislocations. This is accompanied by the breaking of cementite plates into separate weakly misoriented fragments. Then, with an increase in the degree of plastic deformation of the material, due to the pulling of carbon atoms from the crystal lattice of cementite, a change in the structure of the carbide can occur. Recall that this process is possible due to a noticeable difference in the average binding energy of carbon atoms with dislocations (0.6 eV) and with iron atoms in the crystal lattice of cementite (0.4 eV) [121, 150].

A typical image of the rail steel plate structure formed as a result of the implementation of the second mechanism of transformation of the lamellar pearlite structure after the long-term operation on the railway is shown in Figs. 4.33 - 4.35.

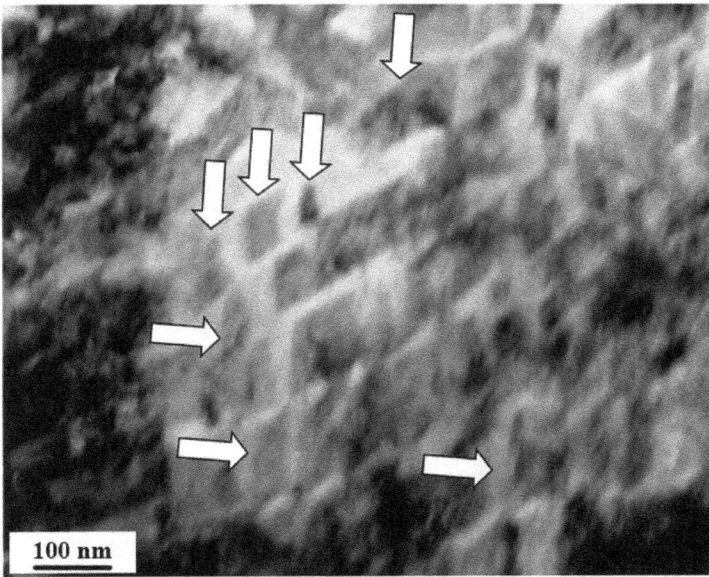

Figure 4.31 TEM image of the structure of a pearlite colony at the stage of cutting cementite plates (indicated by arrows) by gliding dislocations [155, 156].

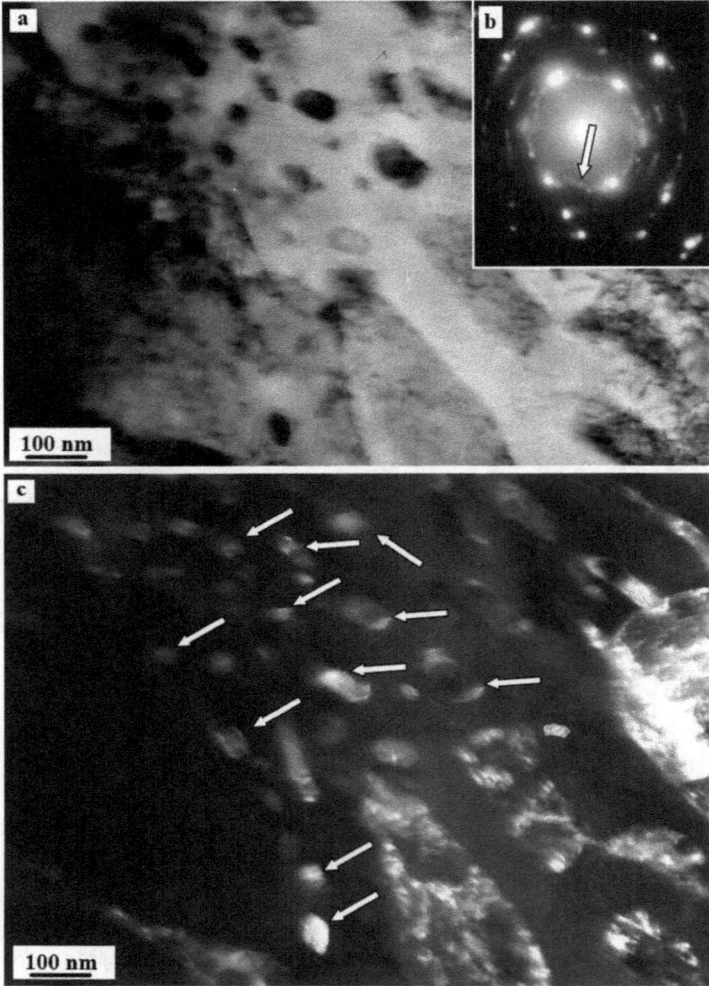

Figure 4.32 *Electron microscopic image of the rolling surface structure; (a) bright field;*
(b) micro-electronogram; (c) dark field obtained in the [012] Fe₃C reflection; on (b) the
arrow indicates the reflex in which the dark field (c) is obtained; on (c) cementite
particles [155, 156].

Fig. 4.33 shows an electron microscopic image of the initial stage of transformation of cementite plates of a pearlite colony, which consists in entangling the plate with sliding dislocations, followed by breaking the cementite plates into separate weakly misoriented fragments.

The second stage of transformation of cementite plates of a pearlite colony, which is realized by the mechanism of dissolution in situ and consists in pulling carbon atoms out of the cementite crystal lattice, is accompanied by a change in the defective carbide substructure. This transformation is caused by the penetration of gliding dislocations from the ferrite crystal lattice into the cementite crystal lattice (Fig. 4.34). Consequently, at this stage of dissolution of cementite plates, the "α-phase / cementite" interphase boundaries play a special role. A coherent or semi-coherent boundary [159] facilitates the penetration of dislocations from α-phase into cementite and vice versa, and thus facilitates the destruction and dissolution of carbide. An incoherent high-angle interphase boundary stabilizes the carbide structure and leaves the possibility only for diffusion mass transfer. That is why cementite plates in a pearlite colony are destroyed, and spherical cementite particles at the grain and subgrain boundaries are preserved.

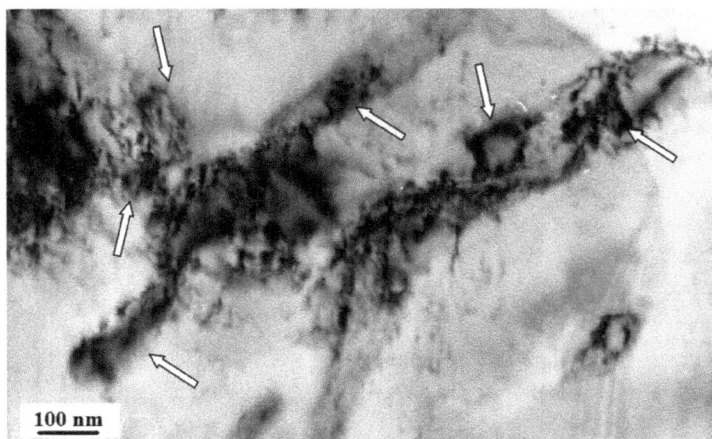

Figure 4.33 TEM image of the structure of a pearlite colony formed during the dissolution of cementite plates by the mechanism in place; (the first stage of the transformation process of cementite plates of a pearlite colony). Arrows indicate fragments in cementite plates [2, 123, 155, 156].

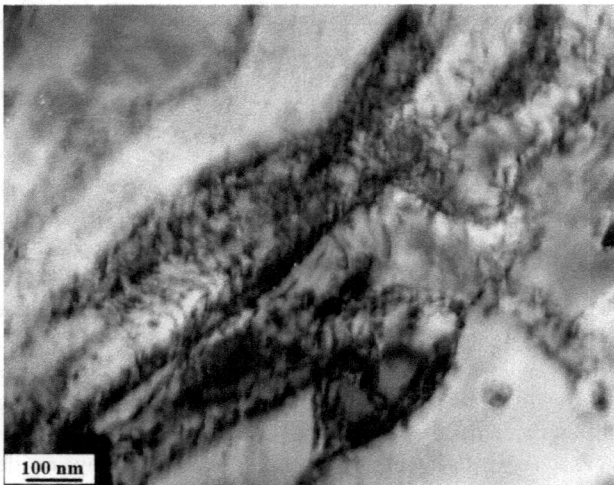

Figure 4.34 *TEM image of the second stage of the process of transformation of cementite plates of a pearlite colony, which is realized by the in situ dissolution mechanism [2, 123, 155, 156].*

At the next stage of cementite dissolution, the entire volume of the material previously occupied by the cementite plate is filled with nanosized particles. A typical image of the resulting structure is shown in Fig. 4.35. Besides, nanosized particles of the carbide phase are also observed in the ferrite matrix filling the interplate space of pearlite colonies. These particles can be carried there in the course of dislocation slip, or which is less likely, formed in the process of deformational decomposition of a solid solution of carbon in the crystal lattice of iron.

The last stage in the evolution of cementite plates, recorded in [121, 150] in the study of drawn pearlitic steel, consists in the formation of a misoriented quasi-strip substructure based on α-phase. Nanosized particles of iron carbide Fe_4C are observed inside and between the bands. No precipitation of other phases was found. Here it is necessary to conclude that in the α-matrix with a dislocation density of $(5 \ldots 6) \times 10^{10}$ cm^{-2}, nanosized particles of Fe_4C carbide are most stable. Cementite and supersaturated (for carbon) carbide $Fe_{20}C_9$ are not detected under these conditions.

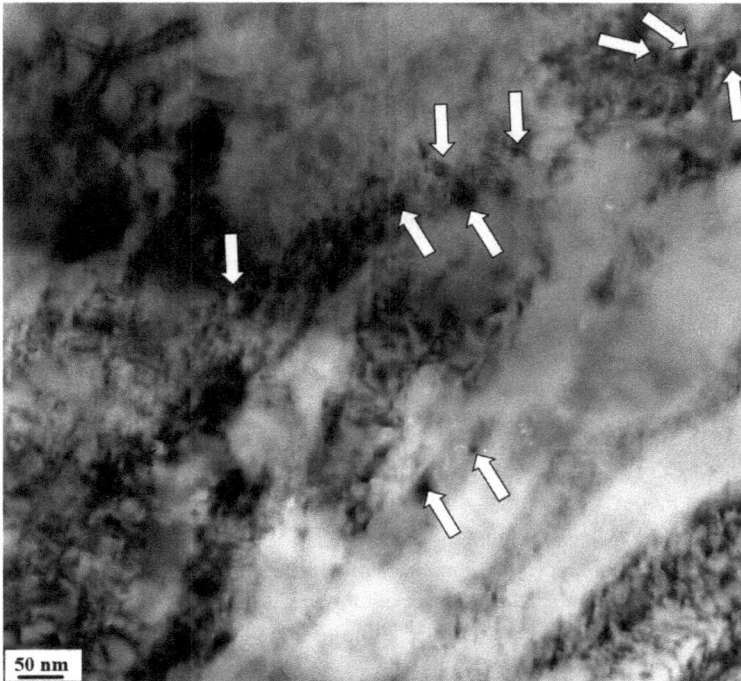

Figure 4.35 *TEM image of the third stage of the process of transformation of cementite plates of a pearlite colony, which is realized by the mechanism of dissolution in place. The arrows indicate nanosized particles of the carbide phase, which are formed in the structure of cementite plates [2, 123, 155, 156].*

Thus, electron microscopic diffraction studies of the evolution of cementite plates of a pearlite colony of low-carbon steel under cold drawing carried out in [121, 150], showed the phase transformations of cementite. The chain of phase transformation of cementite in the process of cold drawing, based on the results of the above studies, is as follows:

$$Fe_3C \rightarrow Fe_3C + Fe_{20}C_9 \rightarrow Fe_{20}C_9 + Fe_4C \rightarrow Fe_4C.$$

Carbon atoms not contained in the Fe_4C carbide particles are concentrated on crystal structure defects (vacancies, dislocations, sub-boundaries, grain boundaries, microcracks) and in a solid solution based on the α-phase.

The "in situ" dissolution process is associated with the mass transfer of at least carbon atoms. It can be carried out through several mechanisms. First, it is the diffusion of atoms at interstitial sites. This mechanism was first proposed in [160] and later confirmed in [158, 161]. It was found [162] that plastic deformation by deformational mass transfer usually represents the directed motion of a large number of vacancies between regions with different signs of internal stresses. Diffusion of carbon atoms over deformation vacancies is the second transfer mechanism. With an increase in the degree of plastic deformation, the total density of dislocations sharply increases, intensive development of the fragmented structure and an increase in the disorientation of fragments occur, and the density of crystal lattice defects in the boundary regions and at grain boundaries increases [2, 123, 163]. This was observed in the study of ferrite-pearlite class 20G2R [164], 70KhGSA [165], and steel 10 [166], as well as pearlite class 9KhF [167-169]. Under the considered conditions, accelerated diffusion of carbon atoms along short circulation paths can occur, i.e. along dislocation tubes, grain boundaries, and fragments [162]. This will be the third transfer mechanism. It is known that the activation energy of diffusion over dislocation cores is much less than over the volume of the material. This mechanism is also included by the authors of works [156, 161] in the list of possible ways of transfer of carbon atoms during the dissolution of cementite. The specific weight of all mechanisms in the process of dissolution of cementite will depend on the structure of the latter, the conditions of deformation, and the degree of alloying of the steel. Dislocations can "lose" carbon atoms, most likely, in areas of the solid solution with significant curvature-torsion of the crystal lattice. These areas are fixed after the entry of carbon there. Otherwise, the process should go in the opposite direction, carbon should pass from the solid solution to dislocations [162].

It should be noted that the considered deformation transformations of the rail steel structure during the operation on the railway did not negatively affect the tribological properties of the rolling surface of the product. Table 4.3 shows the results of wear resistance tests of rails. The tests were carried out on the rolling surface and, for comparison, on a transverse section at a distance of at least 10 mm from the rolling surface.

Table 4.3 *Results of tribological metal tests of rail steel.*

Location	Friction coefficient, μ	Wear factor, 10^{-6}, mm^3/N·m
Rolling surface	0.43	5.1
15 mm from the rolling surface	0.42	5.5

Analyzing the results presented in Table 4.3, it can be noted that the wear resistance of the surface layer of steel after the operation is somewhat ($\approx$1.1 times) higher than the wear resistance of the product volume.

Conclusion

Based on the results of assessing the quality of differentially hardened rails of the R350HT category after the hauled load of 691.8 million gross tons, it was found that (1) in terms of the content of chemical elements, the metal meets the requirements of GOST R 51685-2013 for steel grade E76HF; (2) in terms of mechanical properties, hardness, impact strength, contamination with non-metallic inclusions, macro- and microstructure, the quality of the metal meets the requirements of GOST R 51685-2013 for rails of the R350HT and R350HTNN categories; (3) defects of contact-fatigue origin formed on the rolling surface of the head during the operation have an insignificant degree of development, their depth does not exceed 140 μm [2, 123].

It is shown that, regardless of the direction of analysis (rolling surface, central axis, or fillet, along the radius of curvature), steel is a morphologically diverse material. The main type of structure is perlite lamellar morphology; grains of structurally free ferrite and grains of a ferrite-carbide mixture are present in a substantially smaller amount in steel.

It was found that as a result of the long-term operation of rails under conditions of intense deformation in the surface layer of up to 10 mm thick, a gradient structure is formed. This structure shows itself as a regular change in the scalar and excess dislocation density, the amplitude of curvature-torsion of the crystal lattice of steel, and the degree of deformation transformation of the structure of lamellar pearlite [2, 123].

It is shown that plastic deformation of pearlite steel is accompanied by the simultaneous occurrence of two processes of transformation of the structure and phase composition of lamellar pearlite colonies: (1) cutting of cementite plates and (2) dissolution of cementite plates. The first process, which is carried out by the mechanism of cutting carbide particles and pulling away their fragments, is accompanied only by a change in their linear dimensions and morphology. The second process of destruction of plates of cementite of pearlite colonies is carried out by the escape of carbon atoms from the crystal lattice of cementite at a dislocation, as a result of which a phase transformation of the material is possible with the formation of carbide particles of the composition Fe_4C at the final stage.

5. Carbon Re-Distribution in the Structure of Rail Steel after Long-Term Operation

Iron is practically not used in its pure form in the industry. The basis of the overwhelming number of alloys used in the technology is the iron-carbon system. Iron-carbon alloys (steels and cast irons) are the most important metal alloys of modern technology. The production of iron and steel exceeds the production of all other metals combined by more than ten times. Rail steel of the R65 type belongs to high-carbon low-alloy steels [2, 3, 123]. Carbon steels are iron-based alloys containing up to 2% carbon, as well as manganese, silicon, sulfur, and phosphorus in quantities depending on the smelting method [170]. Carbon is not an accidental impurity, but the most important component of carbon steel, which forms its properties. Machine-building plants receive steel from metallurgical plants in annealed or hot-rolled conditions. The structure of structural steels (hypoeutectoid) consists of ferrite and pearlite, instrumental steels of pearlite, and cementite. With an increase in the carbon content in the steel structure, the amount of cementite, which is a very hard and brittle phase, increases. The hardness of cementite is about ten times the hardness of ferrite (800 HB and 80 HB, respectively) [171]. Therefore, the strength and hardness of steel increase with increasing carbon content, while plasticity and toughness, on the contrary, decrease [172-175].

It is noted in [2, 3, 123] that with an increase in the carbon content to 0.8%, the proportion of pearlite in the structure increases (from 0 to 100%), therefore, both hardness and strength increase. But with a further increase in the carbon content, secondary cementite appears along the boundaries of pearlite grains. In this case, the hardness practically does not increase, and the strength decreases due to the increased fragility of the cementite network [170, 173]. Besides, an increase in the carbon content leads to an increase in the cold brittleness threshold: every tenth of a percent of the concentration of carbon atoms increases t_{50} by about 20°. This means that the steel with 0.4 wt.% C already passes into a brittle state at about 0° C, i.e., it is less reliable in operation [174, 175]. The carbon content also affects all technological properties of steel: the more carbon in the steel, the more difficult is to cut it, it deforms worse (especially in a cold state) and welds worse [172].

Thus, the main components of unalloyed steels are iron and carbon; the main phases (in the most important for modern technology pressure and temperature ranges) are solid solutions of carbon in the γ-modification (γ-phase, austenite) and α-modification (α-phase, ferrite) iron, as well as iron carbide - cementite (Fe_3C) [2, 3, 123].

The atomic (ionic) radius of carbon (0.077 nm) is small in comparison with the atomic radius of iron (0.126 nm); therefore, carbon dissolved in iron forms solid interstitial solutions [173]. The locations of carbon atoms in the crystal lattice of iron are octahedral and tetrahedral interstices [173, 176, 177]. Octahedral interstices are formed by six metal atoms, the centers of the imitating spheres of which are located at the vertices of the octahedron (Fig. 5.1, top row). Tetrahedral interstices are formed by four metal atoms located at the vertices of the tetrahedron (Fig. 5.1, bottom row). In this case, the volume of an octahedral polyhedron is significantly larger than that of a tetrahedral one [2, 3, 123].

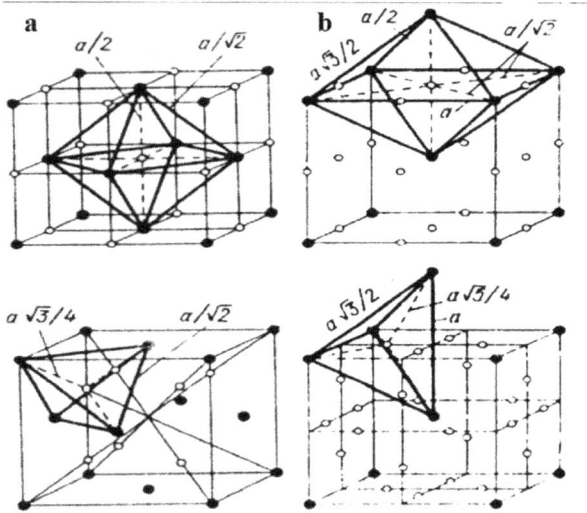

Figure 5.1 *Octahedral (top row) and tetrahedral (bottom row) interstices in (a) - fcc and (b) - bcc crystal lattices (● are metal atoms; ○ are interstitial positions) [176].*

In metals with an fcc structure (γ-iron), the unit cell of which is formed by four atoms, octahedral interstices with cubic symmetry are located in the center of the unit cell and in the middle of its edges (Fig.5.1, a), while tetrahedral interstices are located on spatial diagonals (at a distance of one-quarter of their length from the top) (Fig. 5.1, b). The number of octahedral interstitial positions per unit cell is $n_6 = 4$, and the number of tetrahedral positions $n_4 = 8$. The dimensions of the edges of the corresponding polyhedra are

$$\ell_{4,6} = \frac{a}{\sqrt{2}}$$

(a is the period of the cubic lattice), and the radii of the spheres inscribed in them (within the framework of the model of "hard" spheres) are $r_6 = 0.414R$ and $r_4 = 0.225R$ (where R is the radius of the metal atom). Therefore, for the crystal lattice of γ-iron r_6 (Fe) = 0.414 x 0.126 = 0.052 nm; r_4 (Fe) = 0.225 x 0.126 = 0.028 nm [2, 3, 123].

There are six octahedral ($n_6 = 6$) and twelve tetrahedral ($n_4 = 12$) internodes per unit cell of bcc metals (α-iron) containing two atoms. Octahedral internodes are located in the middle of the edges and faces of the unit cell (Fig. 5.1, b, top row). The polyhedron corresponding to these positions with an edge

$$\ell_6 = \frac{a\sqrt{3}}{2}$$

somewhat flattened, due to which two of its spatial diagonals are $\frac{2a}{\sqrt{2}}$ long and the third one equals a (Fig. 5.1, b, top row). The centers of the tetrahedral internodes are located in the middle between the centers of the octahedral positions (Fig. 5.1, b, bottom row). Tetrahedron edge size

$$\ell_4 = \frac{a\sqrt{3}}{2}$$

The radius of a sphere inscribed in an octahedral polyhedron is $r_6 = 0.154R$; in the case of a tetrahedral polyhedron, $r_4 = 0.291R$. Consequently, for the crystal lattice of α-iron r_6 (Fe) = 0.154 x 0.126 = 0.019 nm; r_4 (Fe) = 0.291 x 0.126 = 0.037 nm.

Based on the geometric constructions and the estimates presented in [2, 3, 123], it follows that the solubility of carbon in crystal lattices based on α- and γ-iron should differ by orders of magnitude. As a result of numerous studies, it has been established that the maximum solubility of carbon in α-iron, achieved at a eutectoid temperature of 727 °C, varies within 0.01 - 0.095 wt.%; at room temperature, it does not exceed 0.006 wt% [170, 178]. The solubility of carbon in γ-iron increases with increasing temperature and reaches a maximum value of 2.14 wt.% at a eutectic temperature of 1147 °C. In the fcc lattice of γ-iron, carbon atoms are located in octahedral pores in the middle of the ribs and in the center of the γ-Fe unit cell (Fig. 5.2) [170, 173, 177].

It should be noted that iron-carbon alloys are of practical importance, the carbon concentration in which does not exceed 6.687 wt.% (Fe-6.687 %C corresponds to the chemical compound Fe_3C). Alloys, the concentration of carbon in which does not exceed 2.14 wt. %C are called steels, and at higher concentrations are cast irons [2, 3, 123].

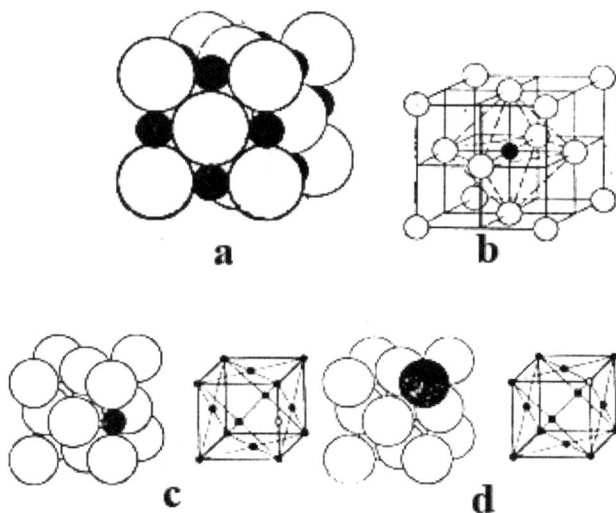

Figure 5.2 *Unit cell of austenite (diagram). Carbon atom (●) in the position of incorporation; (a) crystal lattice when all pores are filled; (b, d) are partial filling of pores, different positions of octahedral internodes [179, 180].*

Iron carbide (cementite), a chemical compound of carbon with iron (Fe₃C), contains 6.67 wt. %C. Cementite has an orthorhombic unit cell with twelve iron atoms and four carbon atoms. The crystalline structure of cementite is complex. Several variants of its structure are discussed. According to one of the early models, carbon is located in the octahedral pores of the iron crystal lattice. One of the most successful graphic methods of depicting this variant of the crystalline structure of cementite is shown in Fig. 5.3, a [170].

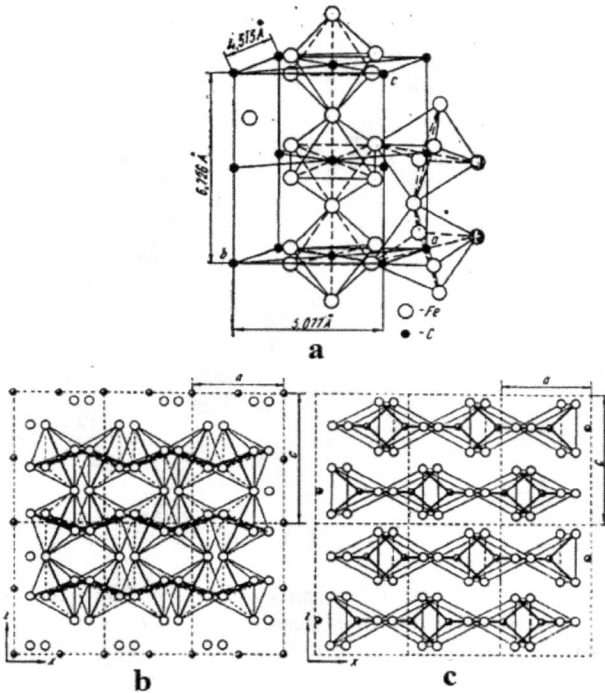

Figure 5.3 *Crystalline structure of cementite; (a) three-dimensional image of the crystal lattice; (b, c) projections of crystal structure options on a plane (b - octahedral, c - prismatic model). Open circles - Fe atoms; dark - C) [170, 177, 181].*

Later studies using neutron diffraction methods led to the creation of a prismatic version of the structure of the cementite crystal lattice (Fig. 5.3, b, c) [177, 181]. In this case, carbon atoms are located in tetrahedral pores [2, 3, 123].

Different solubility of carbon in polymorphic modifications of iron, the very phenomenon of polymorphism, and the possibility of separating several phases from a supersaturated solid solution determine the variety of phase transformations that occur in iron-carbon alloys when they are cooled after solidification. In this case, the most important role in the formation of the final structure of the steel is played by polymorphic recrystallization $\gamma \Rightarrow \alpha$ (Fe), which underlies the eutectoid decomposition of austenite [170, 173, 180-189]. Since carbon is relatively well soluble in the γ-phase and practically insoluble in the α-phase, the final result of the transformation is mainly determined by the possibility of diffusional redistribution of carbon atoms. Under conditions of high diffusion mobility of carbon, austenite undergoes eutectoid decomposition into ferrite and graphite (stable variant, redistribution of iron atoms is well developed) or ferrite and cementite (metastable variant, conditions of weak diffusional redistribution of iron atoms). The latter variant leads to the formation of a finely differentiated ferrite-cementite structure called pearlite because of the pearlescent color of thin sections etched with a standard nitric acid reagent [170]. Pearlite, sorbitol, and troostite are distinguished depending on the dispersion of the structure [190].

It was noted in [2, 3, 123] that suppression of carbon diffusion by high-speed cooling (quenching) leads to the formation of a martensitic structure. Martensite is the main structural component of hardened steel; is a supersaturated solid solution of carbon in α-iron of the same concentration as that of the initial austenite [170, 173, 180-189]. The martensitic transformation is diffusion-free (shear), in which the change in the mutual arrangement of the atoms constituting the crystal occurs through their ordered displacement, and the relative displacements of neighboring atoms are small in comparison with the interatomic distance. A necessary condition for martensitic transformation, which develops through the formation and growth of regions of a more stable phase in a metastable phase, is the maintenance of an ordered contact between the initial and final phases. The ordered structure of interphase boundaries with a small barrier for a homogeneous phase transition ensures their low energy and high mobility. As a consequence, the excess energy required for the nucleation of new phase crystals (martensitic crystals) is small. Therefore, the nucleation of martensite crystals occurs at a high rate and may not require thermal fluctuations. Due to the effect of the formed phase on the initial phase, the energy barrier for the displacement of the phase boundary is significantly less than for a homogeneous transition; at small deviations from equilibrium it disappears, the crystal grows at a speed of the order of sound and without thermal activation. Martensitic transformations have been found in many crystalline materials: pure metals, numerous alloys, ionic, covalent, and molecular crystals. Martensitic transformations in iron-based alloys, in particular, in connection with the hardening of steel are the most extensively studied [2, 3, 123].

In carbon steels in the range of 500–250 °C, a bainitic or intermediate transformation occurs, which is an intermediate one between pearlite and martensitic [153, 170, 173, 191]. During the transformation, a ferrite-cementite mixture is also formed, but the carbide particles do not have a lamellar structure and are very dispersed (visible only in an electron microscope). "Upper" and "lower" bainite, formed respectively in the upper

(550-4000) and lower (400-250°) parts of the intermediate temperature range are distinguished.

It was noted in [2, 3, 123] that, in contrast to pearlite, bainitic transformation is a shear diffusion transformation; the rearrangement of the crystal lattice occurs due to the cooperative displacement of iron atoms at distances less than interatomic in the diffusion displacement of carbon atoms. Upper and lower bainite differ from each other in structure and strength characteristics. The first has a feathery structure, relatively low strength and plasticity; the second is an acicular structure (close to martensite), high strength and plasticity.

The main difference between bainite and pearlite structures is the carbon content in ferrite. At high temperatures, carbon has time to release from the solution and ferrite contains about 0.01-0.02 % C. At low temperatures (about 500-250 °C), the diffusion rates are low, carbon does not have time to completely release from the solution, therefore ferrite contains approximately 0.1 wt. % C (400 °C) and even about 0.2 wt. % C (300 °C) [2, 3, 123].

Thus, the results presented above indicate that the structure and properties of carbon-containing steels are significantly affected not only by the concentration but also by the arrangement of carbon atoms in the material structure. Let us consider the issue related to the distribution of carbon atoms in steel in more detail.

We noted [2, 3, 123] that in the first chapter and several works [121, 152, 153, 173, 192-201] it was shown that carbon in the steel structure can be found in a solid solution based on α- and γ-iron (at the position of interstitial elements), at dislocations (in the form of Cottrell and Maxwell atmospheres), at interfacial (carbide/matrix) and intraphase (grain boundaries, packets, and crystals of the packet and lamellar martensite) boundaries, in particles of the carbide phase. The amount of carbon in solid solutions based on α- and γ-iron is usually estimated from the relative change in the crystal lattice parameter of these phases [202–204]. Estimates of the amount of carbon in carbide particles are carried out based on the chemical composition of the carbide, the type of crystal lattice, and the volume fraction of the particles of the carbide phase in steel. For cementite (assuming a stoichiometric composition), a similar calculation was performed in [205]. Estimation of the amount of carbon located on defects (dislocations and interfaces) is the most difficult moment and is difficult to determine directly experimentally. Quite often this situation is resolved by using indirect methods (for example, methods of internal friction and micro-X-ray spectral analysis) [192-197, 201], as well as theoretical estimates. We noted in [2, 3, 123] that the most complete analysis of the redistribution of carbon in unalloyed steels, depending on the tempering temperature, was carried out in [193-197], and in [206, 207]

Structure and Properties of Lengthy Rails after Extreme Long-Term Operation
Materials Research Foundations **106** (2021)

Materials Research Forum LLC
https://doi.org/10.21741/9781644901472

the case of alloyed steels (quenching and low-temperature tempering state) was considered. In [208], the results of quantitative studies of the structural-phase state of hardened steel 38KhN3MFA are presented, the locations of carbon are identified, and its redistribution is analyzed depending on the austenitization temperature. In [152], similar estimates were made for hardened steel subjected to various degrees of plastic deformation by uniaxial compression; in [153, 209] for steel with bainitic structure; in [121] for carbon steels subjected to various types of loading.

This chapter presents the results of a study by the methods of X-ray phase analysis, optical and transmission electron diffraction microscopy of the structural-phase state, and distribution of carbon atoms in the rail metal (along the central axis) after the hauled load of 691.8 million gross tons.

It was noted earlier in [2, 3, 123] that the main type of structure of the investigated steel is the grains of lamellar pearlite, the relative content of which in the material is 0.9; the relative content of grains of the ferrite-carbide mixture is 0.05; the rest are grains of structurally free ferrite. The state of the rail metal close to the original steel after the hauled load of 691.8 million gross tons during field tests is obtained at a distance of 10 mm from the rolling surface along the central axis. A typical electron microscopic image of the structure of a given rail metal layer is shown in Figs. 5.5 and 5.6.

As indicated in [2, 123], a distinctive feature of the structure of the metal layer (from the initial metal of rail steel) of the rails located at a distance of 10 mm from the rolling surface, after hauled load of 691.8 million gross tons during field tests, is a large number of bending extinction contours (Figs. 5.4 and 5.5). The presence of bending extinction contours indicates elastic-plastic distortions of the crystal lattice of the material, which may be caused by the mechanical action of the rolling stock on the rail metal during the operation [121, 153].

The operation of the rails, as shown by the studies carried out by the methods of transmission electron diffraction microscopy [2, 123], is accompanied by the simultaneous occurrence of some processes, the main of which are the following. First, an increase in the scalar dislocation density in the range from 2.9×10^{10} cm^{-2} (in a layer located at a distance of 10 mm from the rolling surface) to 3.3×10^{10} cm^{-2} near the rolling surface. In this case, the type of dislocation substructure (chaotically distributed dislocations and dislocation networks) does not change.

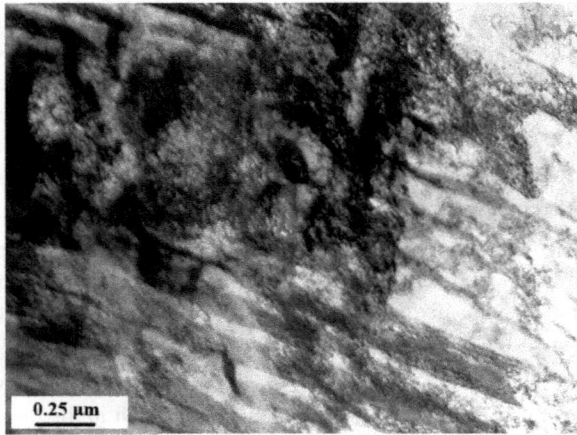

Figure 5.4 *Electron microscopic image of the structure of lamellar pearlite in rail steel after the operation. A layer located at a distance of 10 mm from the rolling surface [2, 123].*

Secondly, the destruction of the lamellar pearlite structure by cutting cementite plates and dissolving them. A typical image of the resulting structure is shown in Fig. 5.6. The destruction of cementite plates is accompanied by the formation of globular particles, the dimensions of which in the surface layer are: transverse 30 nm and longitudinal 50 nm. As the distance from the rolling surface increases, the particle sizes also increase in the layer located at a distance of 10 mm from the rolling surface. The sizes of these particles are 30 x 215 nm. Studies carried out by the methods of dark-field analysis [2, 123] show that the operation of rails is accompanied by the crushing of cementite plates with their subsequent destruction. The reflections of the carbide phase in the microelectron diffraction pattern obtained from such plates have both radial and azimuthal smearing, which may indicate a high level of defectiveness of the cementite crystal lattice, as well as a change in the crystal lattice parameter due to the escape of carbon atoms [113, 210].

Third, simultaneously with the destruction of pearlite colonies, the volume fraction of cementite in the studied steel decreases from 11.2% in a layer located at a depth of 10 mm to 4.9% in the surface layer. This fact indicates the dissolution of cementite plates and the escape of carbon atoms into the crystal lattice of α-iron and onto defects in the crystal structure of steel (dislocations, boundaries, and subboundaries).

Materials Research Forum LLC
https://doi.org/10.21741/9781644901472

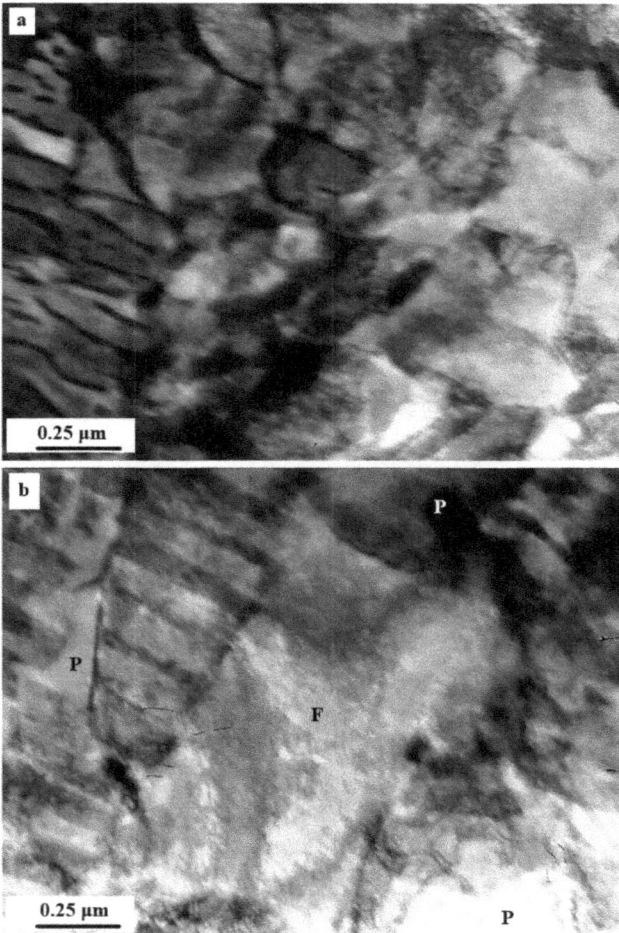

Figure 5.5 *Electron microscopic image of the structure of grains of a ferrite-carbide mixture (a) and structurally free ferrite (b) of rail steel at a distance of 10 mm from the rolling surface along the central axis; on (b) symbol "F" shows ferrite grains; P - pearlite grains [2, 123].*

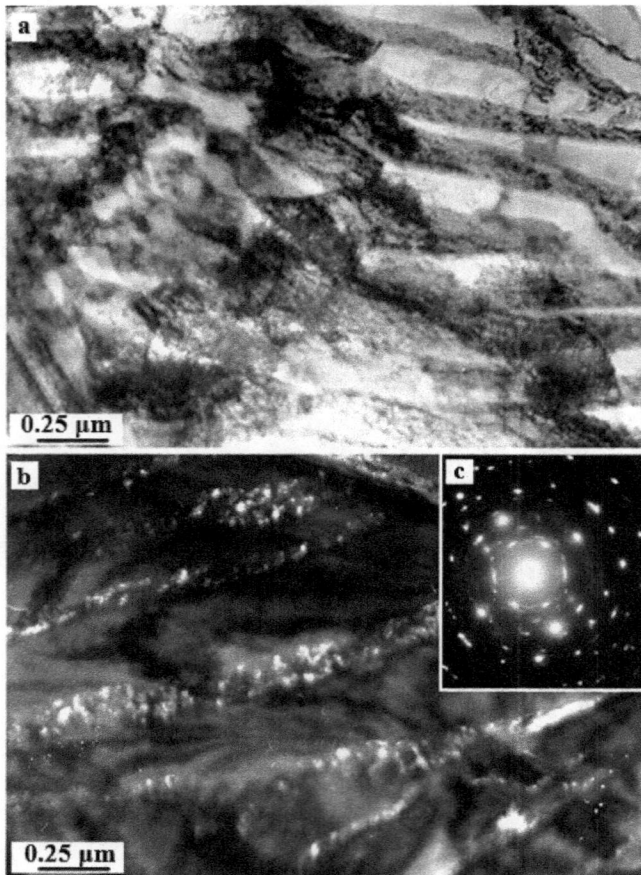

Figure 5.6 *Electron microscopic image of the structure of rail steel of the rolling surface along the central axis; (a) bright field; (b) dark field obtained in the reflection [112] Fe₃C; (c) micro-electronogram, the arrow indicates the reflex in which the dark field was obtained. The surface layer is approximately 2 mm thick [2, 123].*

The concentration of carbon atoms in the crystal lattice of α-iron was determined by X-ray structural analysis. X-ray studies carried out in the work consisted in determining the lattice parameter of the α-phase; the size of the regions of coherent scattering and the magnitude of microstresses. The method for their determination is described in [2, 123].

Structure and Properties of Lengthy Rails after Extreme Long-Term Operation
Materials Research Foundations **106** (2021)

Materials Research Forum LLC
https://doi.org/10.21741/9781644901472

The calculation of the cubic lattice parameter (a_0) was performed, as noted in Chapter 2, using the relation [113, 210, 211]:

$$a_0 = d_{hkl}\sqrt{h^2 + k^2 + l^2}$$

where d_{hkl} is the interplanar distance; h, k, l - plane indices.

The sizes of coherent scattering regions (CSRs) were calculated at small diffraction angles using the Scherrer formula [115]:

$$D_{hkl} = \frac{K \cdot \lambda}{\beta \cdot \cos \theta_{hkl}}$$

where D_{hkl} is the CSR size, K is the coefficient accounting for the shape of the particles, λ is the X-ray radiation wavelength, β is the half-width of the X-ray reflection, and θ_{hkl} is the diffraction angle. The calculation was carried out at $K = 0.94$.

Microdistortions of the α-iron crystal lattice (calculated from the broadening of the reflection using the expression [115]:

$$\frac{\Delta d}{d} = 0.25 \cdot \beta \cdot \cot \theta_{hkl}$$

The previously revealed quantitative regularities of changes in the parameters of the rolling surface structure along the central axis of rail steel after hauled load of 691.8 million gross tons [2, 123] made it possible to carry out studies aimed at analyzing the distribution of carbon atoms in the steel structure.

Estimates of the relative content of carbon atoms on the structural elements of steel were carried out based on the expressions summarized in Table 5.1. The results of the evaluations performed are presented in Table 5.2.

Table 5.1 *On the method of analyzing the distribution of carbon in the steel.*

Carbon locations	Evaluation expressions	Literary source
α- Iron based solid solution	$\Delta C_\alpha = \Delta V_\alpha \dfrac{a_\alpha - a_\alpha^0}{39 \pm 4} \cdot 10^3$ *	[28, 38, 39]
Particles of carbide phases	$\Delta C(Fe_3C) = 0.07*\Delta V_i$	[28, 42, 45]
Defective structure elements	$\Delta C_{\dashv} = C_0 - \Delta C_\alpha - \Delta C(Fe_3C)$	[28, 45]

* Here ΔV_α, ΔV_i are volume fraction of α-Fe and carbide phases, respectively; a_α is the current parameter of the α-phase lattice; $a_\alpha^0 = 0.28668$ nm [178]; $a_\alpha = 0.28782$ nm; C_0 is the average carbon content in the steel.

The estimates have shown that the operation of rail steel is accompanied by a significant redistribution of carbon atoms in the surface layer of the product. If in the initial state the main number of carbon atoms was concentrated in cementite particles, then after the operation of rails, the location of carbon, along with cementite particles, are defects in the crystal structure of steel such as dislocations, grain boundaries, and subgrains. In turn, carbon atoms were found in the surface layer of the steel and in the crystal lattice based on α-iron [212].

Table 5.2 *Distribution of carbon in the structure of rail steel after the hauled load of 691.8 million gross tons.*

Structural elements	Carbon concentration, weight, %		
	Surface	2 mm from the surface	10 mm from the surface
Particles of cementite	0.33	0.71	0.75
Crystal cell α-Fe	0.0284	0.0	0.0
Crystal structure defects	0.3816	0.03	0.0

Thus, the obtained results of a quantitative analysis of the parameters of the steel structure made it possible to trace the redistribution of carbon atoms in the material structure. These results were obtained in the study of the rolling surface of the rail after the hauled load of 691.8 million gross tons. It was found that the operation of rails is accompanied by intensive destruction of cementite particles in the surface layer of steel; carbon atoms that left the crystal lattice of cementite particles are located on defects in the crystal structure of steel (dislocations, grain boundaries, and subgrains). In a layer located at a depth of 2 mm or more, the main location of carbon atoms is cementite particles.

Conclusion

The phase composition, macro- and microdefect structure of the rolling surface metal along the central axis of differentially hardened rails after the operation were studied by X-ray diffraction analysis, optical and electron diffraction microscopy [2, 123]. It is shown that the operation of rails is accompanied by multiple transformations of the steel structure. This is manifested in the formation of microcracks at the macro level; at the microlevel - in an increase in the scalar density of dislocations, the formation of elastic-plastic stress fields, and destruction of cementite plates of pearlite colonies. It is shown that carbon atoms that have left the crystal lattice of cementite particles are located mainly on defects in the crystal structure of steel (dislocations, grain, and subgrain boundaries). In steel, before the operation of the rail, the main location of carbon was particles of the carbide phase (cementite) [212].

6. Physical Nature of Hardening of Metal of Differentially Hardened Rails in the Process of Long-Term Operation

6.1   Evolution of mechanical, tribological properties, macro- and microstructure of the surface layer of metal of differentially hardened rails during long-term operation

To date, a large number of works have been published in which the main attention is paid to the quantitative assessment of the mechanical properties of metals and alloys, including steel, both at the yield point and at the stage of active plastic deformation [2, 3, 152, 153, 213-216]. Significant progress has been achieved in understanding the behavior of the material under various types of external influences, based on the analysis of the elemental and phase composition, the state of the defective substructure and crystal lattice of the main phases, internal stress fields, and their sources (stress concentrators) [2, 3, 213, 214]. Particular attention was focused on the problem of strength, the behavior of which can currently be predicted in many cases with a high degree of reliability, based on knowledge of the alloy composition and its microstructure [2, 3, 152, 153, 213-218]. Estimated relationships that make it possible to reveal the physical nature of hardening of a material are often obtained based on physical models which describe the phenomenon of hardening. However, in some cases, when developing them, empirical or semi-empirical preconditions are used, especially when it is necessary to interpret properties based on an analysis of the behavior of complex structural-phase states [2, 152, 153].

As noted in [2, 3], understanding the hardening mechanisms makes it possible to obtain the optimal combination of material properties necessary for its successful application in practice, to evaluate the factors that control these mechanisms and their role in the formation of the most important properties, such as viscosity, strength, and plasticity. In the problem of the formation of the properties of metallic materials, as a rule, three fundamental types of hardening are distinguished: 1) solid solution hardening (substitutional and interstitial atoms, structural vacancies, short-range and long-range order, antiphase domains, etc.), 2) substructural hardening due to linear and flat defects, 3) multiphase hardening (carbides, nitrides, carbonitrides, inclusions of retained austenite in steels, eutectics, composites, etc.) [2, 3, 152, 153, 213-218].

The monographs [2, 3] show that the operation of products in an industrial environment is accompanied by significant changes in the structure and properties of the material. The long-term operation of rails leads to the evolution of the defective substructure, phase and elemental composition, and, subsequently, to the properties of the surface layer, the thickness of which is determined by many factors [46–49, 121, 149]. Under certain conditions, these changes lead to the failure of the rails [219, 220].

The operation of rails, as shown in [2, 3], is accompanied by a significant change in the structural-phase state of the material, both in the surface layer and in the near-surface volume approximately 10 mm thick. The results of mechanical and tribological tests of differentially hardened rails after long-term operation [2, 3], indicate that the requirements of GOST 51685-2013 for rails of the R350HT category are met. The hardness of the metal changes depending on the distance to the working surface of the product, which is due to the gradient of the phase composition and defective substructure of the material formed during differential hardening.

Tribological tests of the rolling surface metal (the test method is detailed in [2, 3, 122]) showed a slight increase in wear resistance and the coefficient of friction of the rail rolling surface metal relative to the rail metal located at a distance of 15 mm from the rolling surface along the central axis. The parameters of the micro- and macrostructure of the metal of differentially hardened rails after the long-term operation, as noted in [2, 3], meet the requirements of GOST 51085-2013, 1778-70, 8233, 5639-82.

Thus, the long-term operation of rails on the railroad leads to changes in both the microhardness and wear resistance of the surface layer of the product. The revealed changes are due to the corresponding evolution of the phase composition and defect substructure and, consequently, the mechanisms of material hardening.

This chapter is devoted to the analysis of the physical nature of the strength of the rail metal based on the numerical assessment of the hardening mechanisms, carried out according to the results of the structural parameters, phase composition, and defective substructure of the rails after the long-term operation, obtained by optical, scanning, and transmission electron diffraction microscopy (thin foil method). The results of the analysis of the structure and phase composition of the metal of differentially heat-strengthened rails are detailed in [2, 3]. First of all, a separate analysis of the mechanisms of strengthening the metal of the working fillet and the rolling surface will be carried out; further, their comparative analysis will be done.

6.2 Phase composition and defective metal substructure of differentially hardened rails during long-term operation

Using the methods of electron diffraction microscopy of thin foils in the metal transmission of a differentiated heat-strengthened rail of the P65 type, category R350HT, it was shown in [2, 3] that at a distance of 22 mm from the rolling surface along the central axis were identified pearlite grains of lamellar morphology (Fig. 6.1), grains of structurally free ferrite (ferrite grains in which there are no particles of the carbide phase) (Fig. 6.2, where **F** is a ferrite grain and **P** is a pearlite grain of lamellar morphology) and grains of a ferrite-carbide mixture (ferrite grains in which cementite particles are randomly located) (Fig. 6.3).

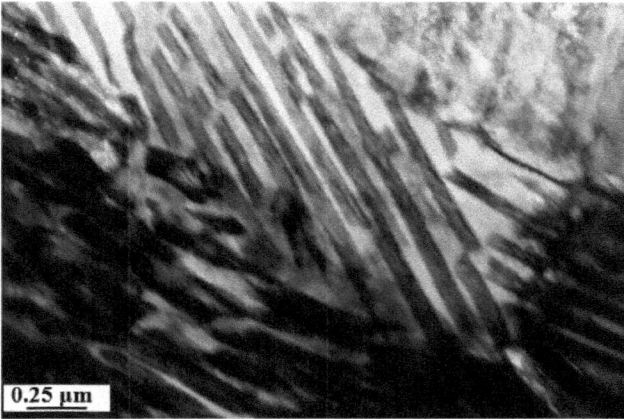

Figure 6.1 TEM image of the structure of lamellar pearlite of rail steel after operation
[2, 3].

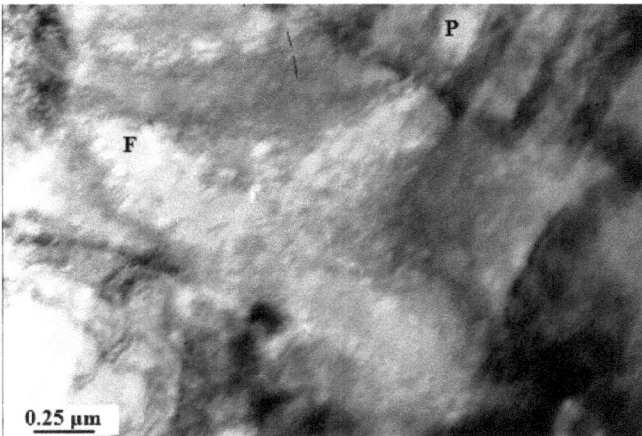

Figure 6.2 TEM image of a grain of structurally free ferrite of rail steel after operation
[2, 3].

After the hauled load of 691.8 million tons [2, 3], the main type of structure of the studied steel is lamellar pearlite grains, the relative content of which in the test material is approximately equal to 0.9; the relative content of grains of the ferrite-carbide mixture is about 0.05; the rest are grains of structurally free ferrite. In the volume of ferrite grains (Fig. 6.4) and the ferrite component of pearlite grains, a dislocation substructure is observed in the form of chaotically distributed dislocations or dislocations that form a network substructure (Fig. 6.5).

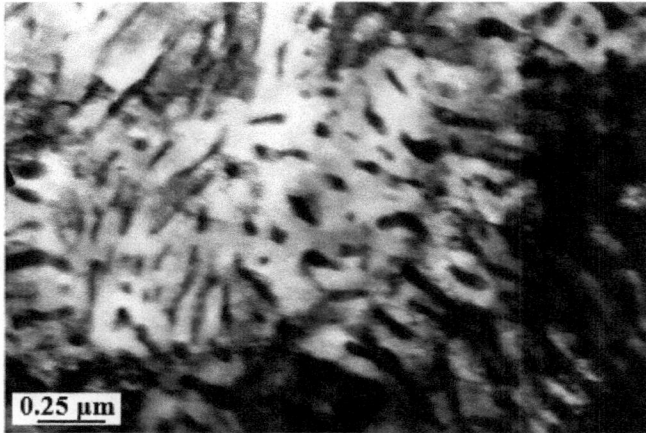

Figure 6.3 TEM image of a grain of a ferrite-carbide mixture of rail steel after operation
[2, 3].

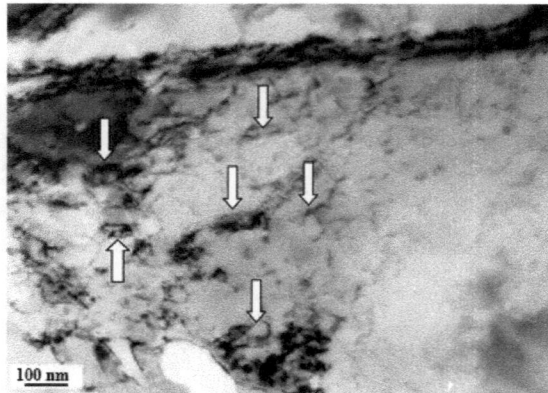

Figure 6.4 Dislocation substructure formed in grains of a ferrite-carbide mixture.
Arrows in the figure indicate particles of iron carbide (cementite) [2, 3].

It is well known [2, 3] that the dislocation substructure of metals and alloys, including rail steel, is quantitatively characterized by the value of the scalar dislocation density, i.e. the total length of dislocation lines per unit volume of the material. The scalar dislocation density was determined by the randomly thrown secant method [113], while the type of dislocation substructure (mesh or chaotic dislocation substructure) was not taken into account. Thus, a certain characteristic of the dislocation substructure averaged over the investigated volume of rail steel was obtained. The research results presented in Fig. 6.6, (a) show that the scalar dislocation density decreases with increasing distance from the surface of the railhead (fillet or rolling surface).

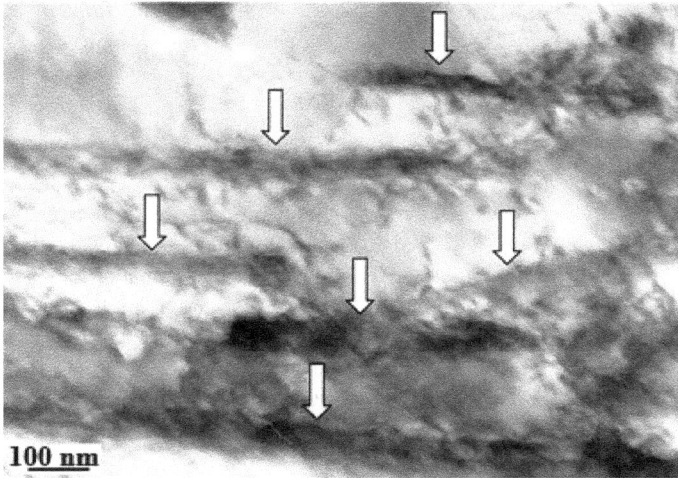

Figure 6.5 *Dislocation substructure formed in the ferritic component of lamellar pearlite grains; arrows indicate cementite plates [2, 3].*

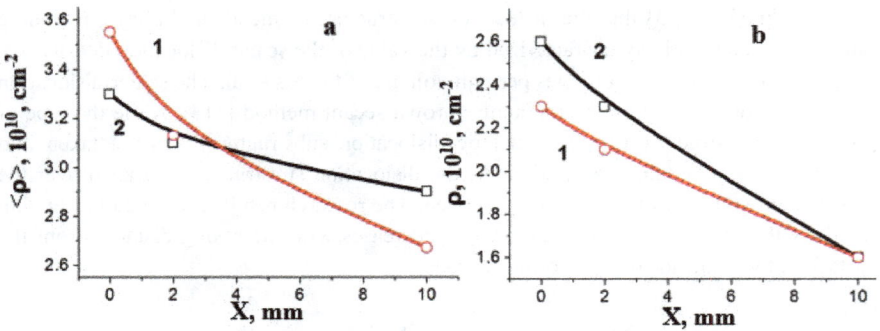

Figure 6.6 *Dependence of the scalar (a) and excess (b) dislocation density on the distance from the working surface of the rails; curve 1 - along the fillet, curve 2 - along the central axis [2, 3].*

Investigation of the structure of rail steel using transmission electron microscopy of thin foils [2, 3] showed the presence of bending extinction contours in electron microscopic images (Fig. 6.7). The presence of bending extinction contours indicates the curvature-torsion of the crystal lattice of the material caused by the formation of internal (long-range) stress fields [112, 113]. The possibility of using electron microscopy of thin foils to study internal stress fields, as noted in [2, 3], was realized by P. Hirsch et al. [112]. The method for practical measurement of internal (long-range) stress fields, based on the analysis of bending extinction contours, has been developed and tested on a wide range of materials (mono-, poly- and nanocrystalline metals and alloys, cermet materials) by a team led by Professor E.V. Kozlov and Professor N.A. Koneva [14, 118, 120, 147, 151, 221, 222]. The studies of the rail metal structure carried out in this work showed that the main sources of curvature-torsion of the crystal lattice (concentrators of internal stress fields) of rail steel are mainly interphase (ferrite/cementite) interfaces.

Structure and Properties of Lengthy Rails after Extreme Long-Term Operation
Materials Research Foundations **106** (2021)

Materials Research Forum LLC
https://doi.org/10.21741/9781644901472

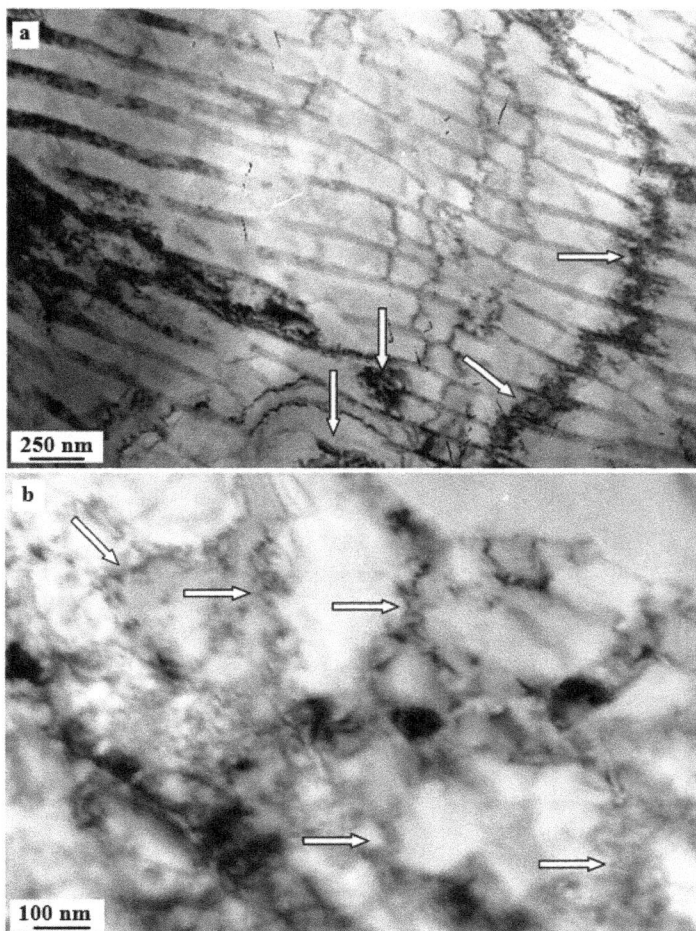

Figure 6.7 *Flexural extinction contours (indicated by arrows) discovered in the structure of lamellar pearlite (a) and grains of a ferrite-carbide mixture (b); the contours are indicated by arrows [2, 3].*

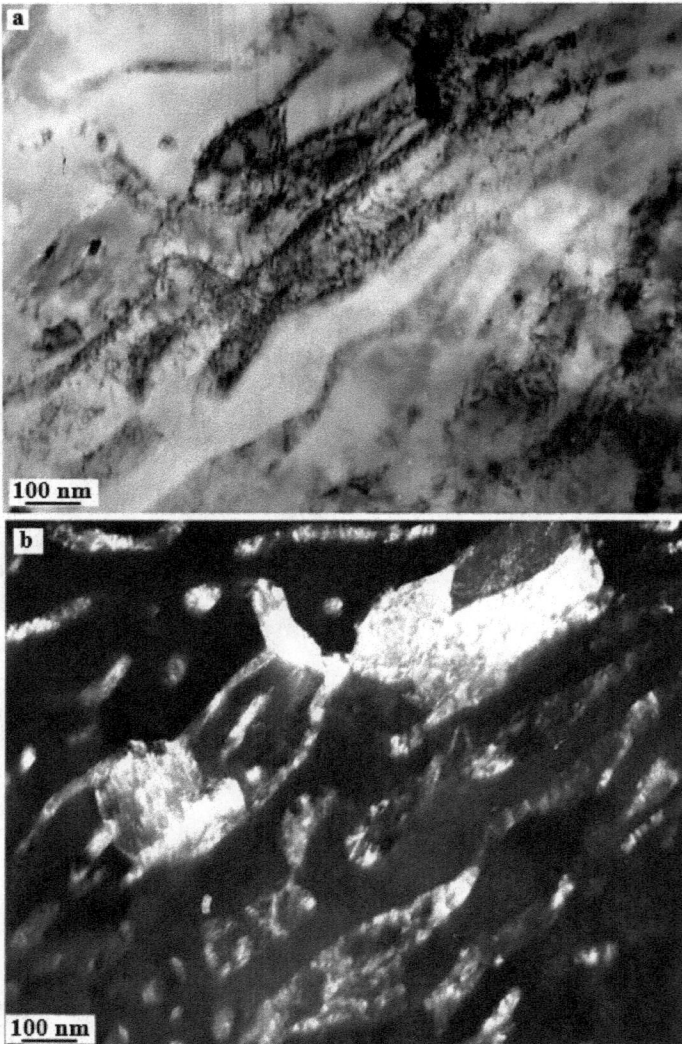

Figure 6.8 *TEM image of the structure of the surface layer of the metal of differentially hardened rails after long-term operation; (a) bright field; (b) dark field obtained in the reflection [012] Fe₃C [2, 3].*

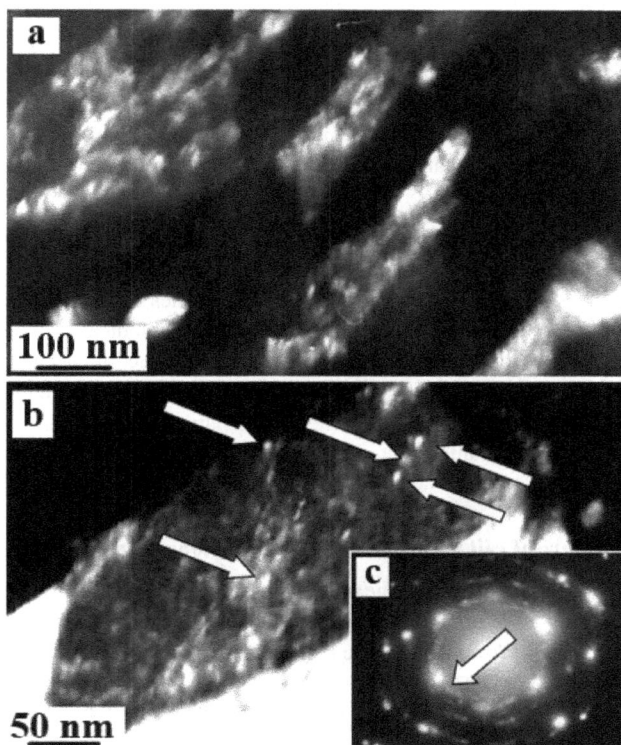

Figure 6.9 *TEM image of the structure of the surface layer of the metal of differentially hardened rails after long-term operation; (a, b) dark fields obtained in the reflection [012] Fe₃C; (c) micro-electronogram, the arrow indicates the reflex in which the dark fields were obtained [2, 3].*

One of the structural characteristics of the curvature-torsion of the crystal lattice of the material is the excess density of dislocations. The dependence of the excess dislocation density on the distance to the working surface of the rails is shown in Fig. 6.7, (b) (the method for estimating the value of the excess density of dislocations will be considered in [2, 3]). Analyzing the results presented in Fig. 6.7, it can be stated that with an increase in the distance from the working surface of the rail (fillet and rolling surface), the value of the scalar and excess dislocation density decreases. It should also be noted that the value of the scalar density for the metal of the working fillet and the metal of the rolling surface is higher than the corresponding values of the excess density of dislocations.

The decarburization of the surface layer of the metal during the operation of rails, discovered on etched thin sections, is also detected indirectly when studying the structure of pearlite grains of steel using transmission diffraction electron microscopy. Fig. 6.8 shows an image of cementite plates of a pearlite colony located in the surface layer of the rolling surface of rails after the operation [2, 3].

Studies carried out by the methods of dark-field analysis (Fig. 6.9) show that the operation of rails is accompanied by crushing of cementite plates (Fig. 6.9, a), followed by their destruction and the formation of nanosized particles of the carbide phase in the interplate space (in a ferrite plate) (Fig. 6.9, b, particles are indicated by arrows) [2, 3]. Reflections of the carbide phase in the microelectron diffraction pattern (Fig. 6.9, c) obtained from such plates have both radial and azimuthal blurring, which may indicate a high level of defectiveness of the cementite crystal lattice, as well as a change in the cementite crystal lattice parameter due to the departure of carbon atoms from its crystal lattice [112, 113, 221].

6.3 Physical nature of rail metal hardening during long-term operation

As noted in [2, 3], the hardening of the surface layer of the metal of rail steel during the long-term operation is multifactorial and is due to the combined action of several physical mechanisms. Let's consider these mechanisms in more detail.
The operation of rails is accompanied by the formation of a fragmented substructure in the surface layer. Strengthening of the material by low-angle boundaries by separating the fragments (substructural hardening, hardening by the boundaries of fragments) can be estimated using the expression [2, 3, 152, 213-216]:

$$\sigma(L) = \sigma_0 + kL^{-m} \qquad\qquad (6.1)$$

where $m = 1$ or $1/2$, L is the average size of the fragments. It was found that at $m = 1$, k varies from 150 to 100 N / m; at $m = 1/2$, k varies from 2×10^{-3} to 10^{-2} Pa · m$^{1/2}$ [2, 3, 152, 213-216].

We used the following values of the parameters included in the equation (6.1), $k = 150$; m = 1 [2, 3]. σ_0 is the frictional stress of the crystal lattice of the material, i.e. stress required for the movement of dislocations in single-phase "pure" single crystals (single crystals that do not contain impurities). The stress σ_0 essentially depends on the purity of the material and the amount of work-hardening. For theoretically pure material $\sigma_0 = 17$ MPa.

The experimentally determined values of σ_0 vary from 27 to 60 MPa [3, 223]. For steels, the value of σ_0 is usually equal to 30 - 40 MPa [3].

It is shown in [2, 3] that in the investigated rail steel, both in the initial state and after the operation, a dislocation substructure with a relatively high scalar dislocation density is revealed. The stress required to maintain plastic deformation, i.e. the flow stress σ_0 required for moving dislocations (deformation carriers) to overcome the forces of interaction with stationary dislocations ("forest" dislocations) is related to the scalar dislocation density as follows [2, 3, 213, 214, 224]:

$$\sigma_\partial = \sigma_0 + \alpha m G b \sqrt{<\rho>} \qquad (6.2)$$

where σ_0 is the flow stress of non-dislocation origin (i.e., caused by other strengthening mechanisms); $<\rho>$ is an average (scalar) dislocation density; m is the Schmid orientation factor; α is a parameter characterizing the value of interdislocation interactions equal to 0.1 ... 0.51 [223, 225]; G is the shear modulus ($\approx$ 80 GPa); b is the Burgers vector of the dislocation (0.25 nm). For steels, taking into account the orientation factor m, it is usually taken $m\alpha \approx 0.5$.

It was shown in [2, 3] that the operation of rails is accompanied by the formation of internal stress fields in steel. The procedure for evaluating the magnitude of internal stress fields is reduced to determining the gradient of curvature-torsion of the crystal lattice χ [14, 151-153, 221]:

$$\chi = \frac{\partial \varphi}{\partial \ell} = \frac{0.017}{h} \qquad (6.3)$$

where h is the transverse dimensions of the bending extinction contour.

According to [2, 3], the value of the excess dislocation density $\rho_\pm$ is related to χ through the Burgers vector of dislocations b

$$\rho_\pm = \frac{1}{b}\frac{\partial \varphi}{\partial \ell} \qquad (6.4)$$

The value of the plastic component of the internal stress fields is estimated based on the ratio [14, 151-153, 221]:

$$\sigma(pl) = m\alpha G b \sqrt{\rho_\pm} \qquad (6.5)$$

The value of the elastic component is estimated based on the ratio [14, 151-153, 221]:

$$\sigma(el) = m\alpha Gbt\chi_{el} \qquad (6.6)$$

where t is the thickness of the foil, taken equal to 200 nm; χ_{el} is the elastic component of the curvature-torsion of the crystal lattice.

The main structural component of rail steel in the initial state and after the operation is lamellar pearlite. The contribution of the pearlite component to steel hardening is estimated by the equation [2, 3]:

$$\sigma(P) = k_y(4.75L)^{-1/2} \times 0.24V(P) \qquad (6.7)$$

where L is the distance between the cementite plates, $V(P)$ is the relative content of pearlite in steel; $k_y = 2 \times 10^{-2}$ Pa·m$^{1/2}$. The performed estimates showed that the contribution of the pearlite component of the structure to the hardening of steel is 165 MPa.

The operation of rails, as noted in [2, 3], is accompanied by the dynamic aging of steel, which leads to the formation of nanosized iron carbide particles in the material. Particles of iron carbide, the sizes of which exceed 5 nm, lose their coherent connection with the crystal lattice of the α-phase [3]. Consequently, the nanosized particles of the carbide phase present in the rail steel, the sizes of which exceed 10 nm, are noncoherent. Incoherent cementite particles are an obstacle to the movement of dislocations, which leads to the strengthening of the material. Steel hardening estimates, taking into account the presence of incoherent particles of the second phase, are carried out using the relation [226]:

$$\sigma_{cp} = M\frac{mG_mb}{2\pi(|\lambda - D|)}\Phi \cdot \ln\left(\left|\frac{\lambda - D}{4b}\right|\right) \qquad (6.8)$$

where λ is the average distance between particles, D is the average particle size, m is an orientation factor equal to 2.75 for bcc materials, $\Phi = 1$ for screw and $\Phi = (1 - \upsilon)^{-1}$ for edge dislocations, M is a parameter accounting for the uneven distribution of particles in the matrix, equal to 0.81 ... 0.85 [225].

The general yield stress of steel in the first approximation based on the principle of additivity, which assumes the independent action of each of the hardening mechanisms of the material, can be represented as a linear sum of the contributions of individual hardening mechanisms [2, 3, 152, 153, 215, 216]:

$$\sigma = \Delta\sigma_0 + \Delta\sigma(L) + \Delta\sigma(\rho) + \Delta\sigma(h) + \Delta\sigma(cp) \qquad (6.9)$$

where $\Delta\sigma_0$ is the contribution due to the friction of the matrix lattice, $\Delta\sigma(L)$ is the contribution due to the intraphase boundaries, $\Delta\sigma(\rho)$ is the contribution due to the dislocation substructure, $\Delta\sigma(h)$ is the contribution due to long-range internal stress fields, $\Delta\sigma(cp)$ is the contribution due to the presence of particles of carbide phases.

Thus, having determined the quantitative characteristics of the steel structure, it is possible, as a first approximation, to analyze the physical mechanisms responsible for the evolution of steel hardness during the operation of the rail, as well as to identify the physical mechanisms of the formation of the rail steel hardness gradient.

6.4 Mechanisms for strengthening the metal of rail steel

It is noted in [2, 3] that the results of a quantitative study of the structure of rail steel, presented in [46-50, 52-56, 141-145, 149, 150, 155, 156, 212, 227], allow us to analyze the mechanisms of hardening and estimate their value depending on the distance to the working surface of the rail using the corresponding expressions for the physical materials science, given above and considered in detail in [2, 3, 152, 153, 213-218, 121]. The calculation took into account the characteristics of the main structural component of the studied steel, i.e. pearlite grains of lamellar morphology. The results of these assessments are shown in Tables 6.1 and 6.2.

Analysis of the results presented in these tables shows that the hardening of rail steel is multifactorial and is determined by the totality of structural components of the material [2, 3]. They are the presence of a dislocation substructure in the material and the degree of its polarization (the formation of internal fields of long-range stresses), the presence of cementite plates of pearlite colonies and nanosized particles of the carbide phase in ferrite plates of pearlite colonies, the formation of a carbon-supersaturated solid solution based on the α-iron crystal lattice. Regardless of the analyzed volume of material (fillet or rolling surface) and the distance to the working surface, the main contribution to the strengthening of the rail steel metal is made by the dislocation substructure, which forms during the operation of the rails.

When analyzing the results obtained, one should bear in mind one more factor of hardening, which is not taken into account in this chapter, the presence of carbon atoms on defects in the crystal lattice of steel (dislocations, grain boundaries, and subgrains). This possibility is indicated by the estimates of the distribution of carbon atoms in the steel structure carried out in [2, 3]. The formation of atmospheres and segregations of carbon atoms at defects in the crystal structure of steel will affect their mobility, i.e. it will strengthen the material.

Table 6.1 *Estimates of the hardening mechanisms of the fillet of rail steel.*

Hardening mechanism	Distance from the surface, mm		
	≈ 0	2	10
Dislocation hardening, σ_∂, MPa	375	350	330
Strengthening by fields of internal stresses: plastic component, σ_{pl}, MPa	300	270	230
elastic component, σ_{el}, MPa	20	30	0
Strengthening with pearlite component, $\sigma(P)$, MPa	48	115	165
Strengthening with cementite particles, σ_{cp}, MPa	67	0	0
Solid solution hardening, $\sigma_{(C)}$, MPa	133	0	0
Additive summation, MPa	943	765	725

Table 6.2 *Estimates of the hardening mechanisms of the rolling surface of rail steel.*

Hardening mechanism	Distance from the surface, mm		
	0	2	10
Dislocation hardening, σ_∂, MPa	363	356	340
Strengthening by fields of internal stresses: plastic component, σ_{pl}, MPa	322	302	253
elastic component, σ_{el}, MPa	34	49	21
Strengthening with pearlite component, $\sigma(P)$, MPa	41	140	165
Strengthening with cementite particles, σ_{cp}, MPa	113	0	0
Solid solution hardening, $\sigma_{(C)}$, MPa	133	0	0
Additive summation, MPa	1006	847	779

Conclusion

Mechanical tests were performed, a comparative layer-by-layer quantitative analysis of the defective substructure and phase composition of the metal of differentially hardened rails after long-term operation was carried out. It is shown that in terms of the level of mechanical properties in tension, the metal of the rails meets the requirements of the standard for rails of the R350HT and R350HTNN categories; the hardness of the rail under investigation meets the requirements of GOST R 51685-2013 for rails of the R350HT category [2, 3].

The mechanisms of strengthening the rail metal at the yield point were considered. Theoretical estimates of the rail metal yield strength are carried out. The multifactorial nature of steel hardening is discovered, which is due, first, to hardening by particles of the carbide phase located in the volume of fragments and on elements of the dislocation substructure (dispersion hardening); secondly, strengthening due to the formation of a pearlite structure; third, strengthening due to the formation of a dislocation substructure; fourthly, hardening introduced by internal stress fields caused by the incompatibility of deformation of the crystal lattices of the structural components of the α-phase and particles of the carbide phase, and, fifthly, strengthening as a result of the formation of a solid solution of carbon in the crystal lattice of α-iron. It is shown that, regardless of the analyzed volume of material (fillet or rolling surface) and the distance to the working surface, the main contribution to the strengthening of the rail steel metal was made by the dislocation substructure, which forms during the operation of the rails. It has been established that the operation of rails is accompanied by the strengthening of the surface layer of the metal. As a result, the strength of the surface layer of steel becomes approximately 1.3 times higher than the strength of the volume of the rails.

7. Structural and Phase State of Rails after the Hauled Load of 1411 Million Gross Tons

7.1 Structure of rail metal at macro and micro levels

As noted in Chapter 2, rails of category R350HT of melt 26891 were removed from the track of the Experimental Ring in the town of Shcherbinka after the hauled load of 1411 million gross tons.

The appearance of the investigated section of the rail is shown in Fig. 7.1.

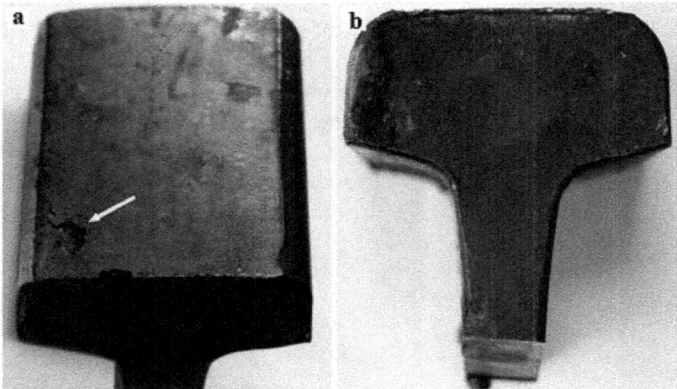

Figure 7.1 *External view of the investigated fragment of the R350HT category rail after the hauled load of 1,411 million gross tons. The arrow indicates the area of chipping of the rail metal on the side of the working fillet.*

The rolling surface of the rail test head has a smoothed and shiny appearance, with a slight shift in wear to one of the fillets (Fig. 7.1, a). In the zone of the working fillet, contact fatigue cracks are observed, located almost at right angles to the rolling axis (Fig. 7.1, b), and small chipping of about 7 x 10 mm in size (Fig. 7.1, a, indicated by an arrow) closer to the end on the indicated side. On the opposite side, there is a slight influx of metal on the surface of the fillet.

The results of the verification chemical analysis of the sample metal are presented in Table 7.1.

Table 7.1 *The chemical composition of the metal rail after hauled load of 1411 million gross tons.*

Chemical analysis	The content of chemical elements, %										
	C	Mn	Si	P	S	Cr	Ni	Cu	V	Al	Ti
Test	0.72	0.77	0.61	0.010	0.009	0.42	0.07	0.14	0.038	0.003	0.003
Requirements TU 0921-276-01124323-2012 for steel grade E76HF	0.71 - 0.82	0.75- 1.25	0.25- 0.60	Not more		0.20- 0.80	Σ not more 0,27 %		0.03- 0.15	Not more	
				0.020	0.020		0.20	0.20		0.004	0.025

As may be seen from Table 7.1, the chemical composition of the rail sample metal meets the requirements of TU 0921-276-01124323-2012 for rails of category R350HT.

Macrostructure

The macrostructure of the metal was obtained by deep etching in a 50% hot aqueous solution of hydrochloric acid on an incomplete transverse template (head, neck) (Fig. 7.2).

Figure 7.2 *Macrostructure of the metal of the investigated rail fragment used to assess the axial segregation, point heterogeneity, segregation bands, and cracks.*

The assessment of the macrostructure was carried out following RD 14-2R-5-2004 "Classifier of defects in the macrostructure of rails rolled from continuous cast steel billets". The macrostructure of the metal of the test sample according to axial segregation, point heterogeneity, segregation stripes, and cracks is assessed satisfactorily (Fig. 7.2). No internal defects or discontinuities were found on the templates. A darker etching area is observed from the rolling surface, the formation of which is associated with the deformation processes of the metal that occur during the operation.

Microstructure

The metal microstructure was studied on thin sections cut from the upper part of the head (fillets and rolling surface) before and after etching in a 4% alcohol solution of nitric acid. When examining un-etched sections by an optical microscope, which were cut from the sample head from the surface of the working fillet at the site of surface contact fatigue cracks, branched discontinuities were discovered, passing at an acute angle to the surface to a depth of 1.09 mm (Fig. 7.3).

After etching, a significantly deformed structure was discovered in the discontinuity zone (Fig. 7.4).

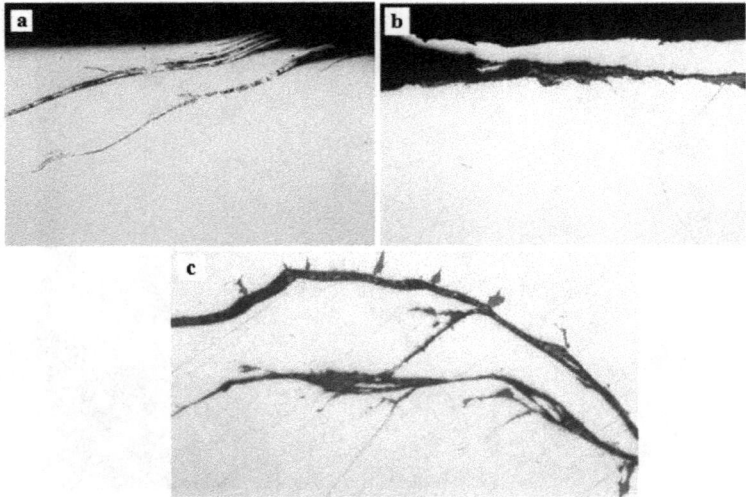

Figure 7.3 *Branched discontinuities detected in the head of a rail fragment from the surface of the working fillet at the site of surface cracks of contact fatigue; (a, b) x50 magnification; (c) x500. Optical microscopy of non-etched thin sections.*

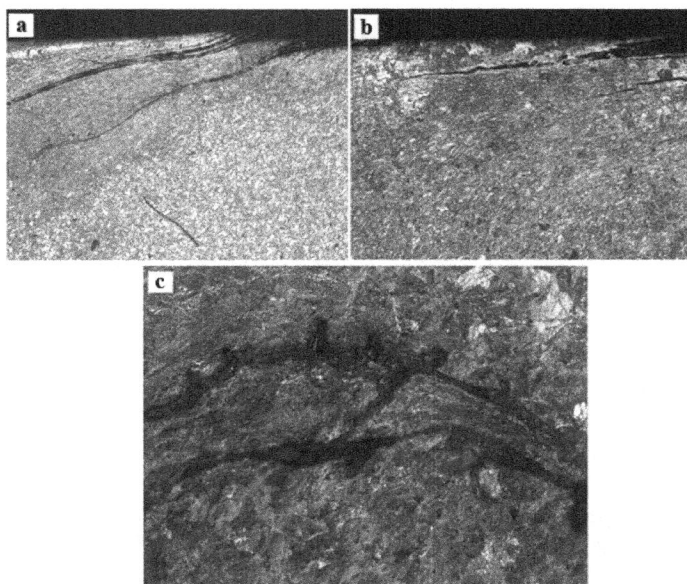

Figure 7.4 *The structure of the rail metal in the area of a branched discontinuity detected in the head of a rail fragment from the surface of the working fillet at the site of surface cracks of contact fatigue; (a) x50 magnification; (b) x100; (c)- x500. Optical microscopy of etched sections.*

There are single small discontinuities up to 0.03 mm deep (Fig. 7.5) on the thin sections cut from the rolling surface of the head. The depth of the deformed layer from the rolling surface, which is clearly visible on etched transverse sections, is insignificant and varies within 0.035 mm (Fig. 7.6). The microstructure in the sample head is formed by grains of finely dispersed lamellar pearlite with small grains of excess ferrite, estimated at 1.5 points of the scale No. 7 of GOST 8233 (Fig. 7.7). Bainite is absent in the microstructure of the sample metal.

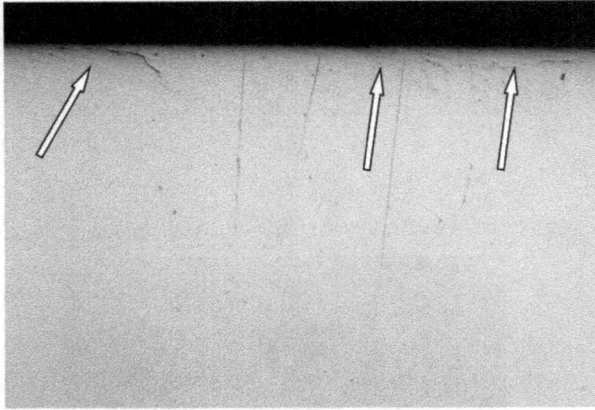

Figure 7.5 *Metal structure of the rolling surface of the head. Optical microscopy of non-etched thin sections, x200. Arrows indicate discontinuities up to 0.03 mm deep.*

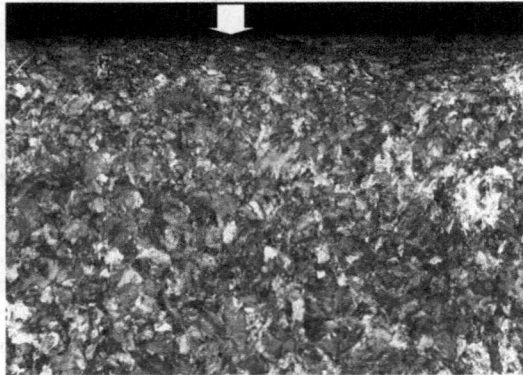

Figure 7.6 *Metal structure of the rolling surface of the head. The arrow indicates the deformed surface layer. Optical microscopy of the etched section, x200.*

Electron microscopic examination of microstructure parameters

For a deeper assessment of the microstructure, a study was carried out on thin sections from the fillet and the rolling surface of the head at a depth of 0.5-1.0 mm using a scanning electron microscope. It has been established that the microstructure of the rail metal is represented by highly dispersed pearlite with the presence of small areas of structurally free ferrite (Fig. 7.8, a). In addition to regular colonies (with the presence of a regular structure of cementite plates), there are quite a few colonies with deformed, fractured cementite plates (Fig. 7.8, b) in the pearlite structure. There are also areas of degenerate pearlite.

Figure 7.7 *Microstructure of metal in the head of a sample of rail steel, shown by optical microscopy; etched thin section. Metal studies were carried out at a depth of 0.5-1.0 mm, x650.*

Estimation of the interplate distance

The results of a quantitative assessment of the microstructure of the metal of steel rails are shown in Table 7.2. Analyzing the results presented in Table. 7.2, a more dispersed structure of the pearlite of the rolling surface can be noted relative to the structure of the fillet pearlite.

Structure and Properties of Lengthy Rails after Extreme Long-Term Operation
Materials Research Foundations **106** (2021)

Materials Research Forum LLC
https://doi.org/10.21741/9781644901472

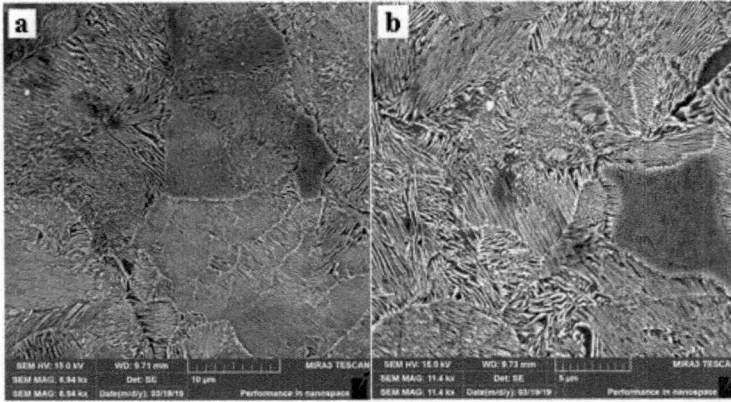

Figure 7.8 *The structure of the railhead metal, revealed by scanning electron microscopy of the etched section. Metal studies were carried out at a depth of 0.5-1.0 mm.*

Table 7.2 *Quantitative characteristics of the structure of the metal of the railhead.*

smelting number	Rail index	IPD, μm			Size of pearlite colonies, μm			Grain diameter, μm		
		min	max	**avg.**	min	max	**avg.**	min	max	**avg.**
26891	Fillet	0.073	0.256	**0.132**	2.711	12.157	**6.17**	15.042	51.169	**29.8**
	RS	0.073	0.225	**0.125**	2.634	10.731	**5.6**	-	-	-

The distribution according to the estimated inter-plate distance (IPD) is shown in Figs. 7.9 and 7.10. It can be concluded from the given distribution in the head zone, that the main level of IPD, which is from 71% to 84%, both in the fillet zone and from the rolling surface, is located in the range 0.10 - 0.15 μm and is at a comparable level. Moreover, as can be seen from the results presented in Table. 7.2, the average value of the IPD from the rolling surface (RS) has a lower value than in the fillet zone.

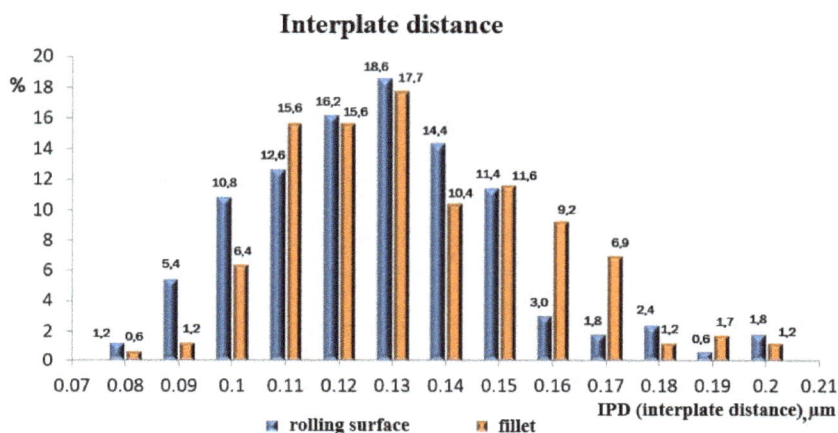

Figure 7.9 *Distribution of the inter-plate distance in the railhead.*

Estimation of the size of pearlite colonies.

The distribution of the estimated size of pearlite colonies (SPC) is shown in Figs. 7.11 and 7.12. From the distributions shown in Fig. 7.12, it can be seen that the main SPC level (ranging from 87.5% to 91.3%) in the fillet zone and from the rolling surface corresponds to the range of 4 - 8 μm. The average value of the size of pearlite colonies in the railhead is 5.6 - 6.2 μm, while the average size of the pearlite colonies at the rolling surface of the head is less than the average size of the pearlite colonies in the fillet zone (Table 7.2).

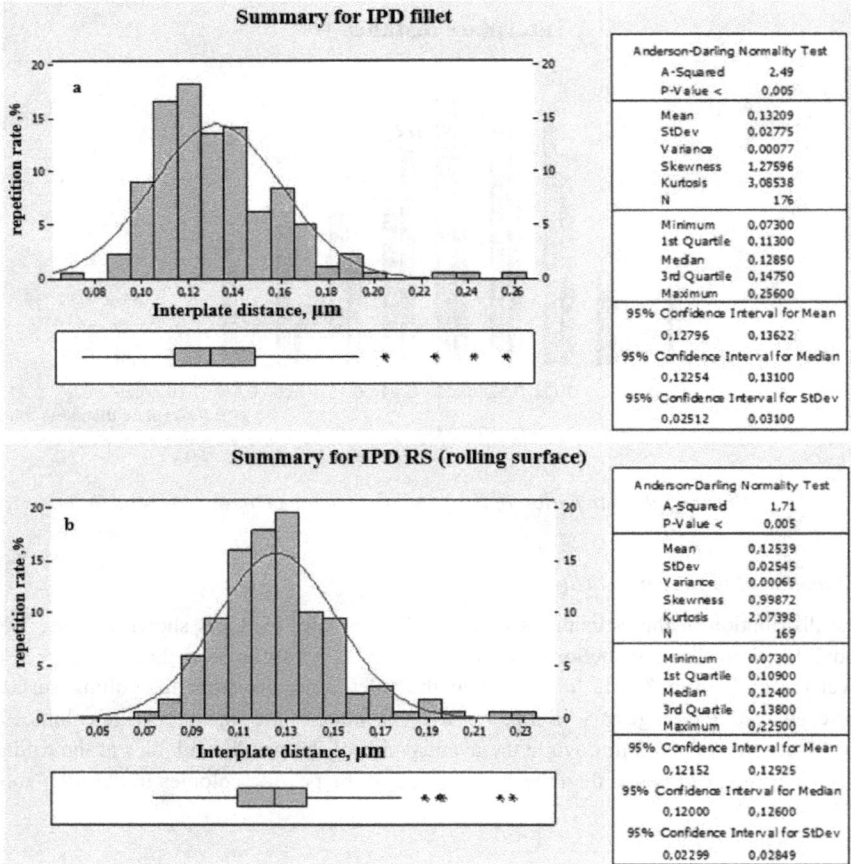

Figure 7.10 *Distribution of the inter-plate distance in the recess zone (a) and the rolling surface zone (b).*

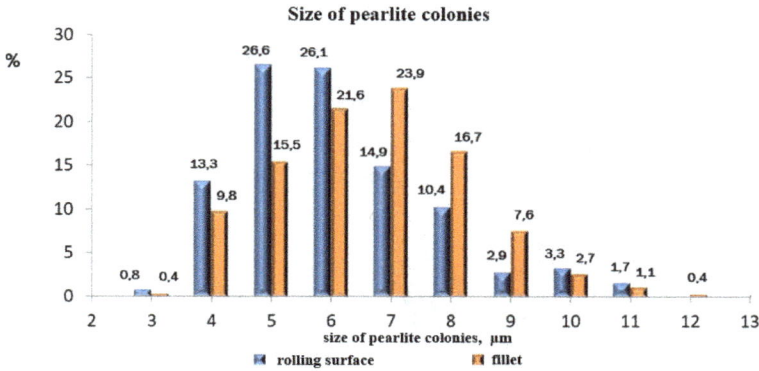

Figure 7.11 *Distribution of pearlite colonies by size in the railhead.*

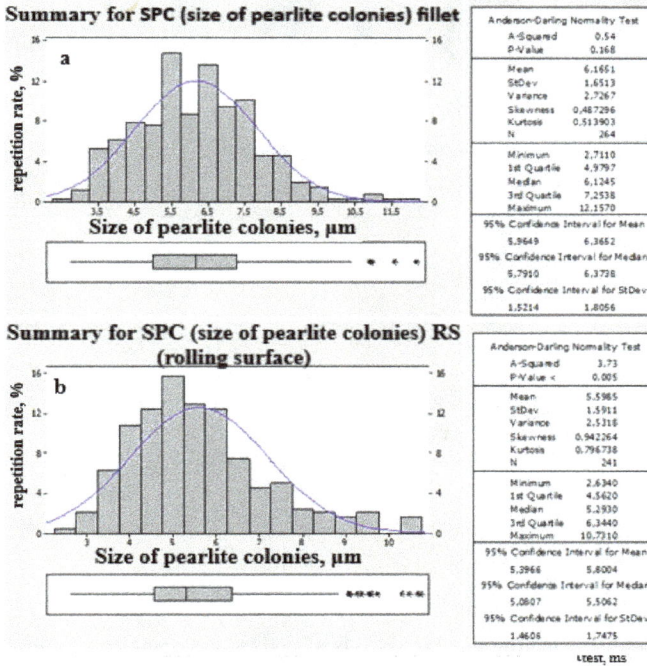

Figure 7.12 *Size distribution of pearlite colonies in the fillet zone (a) and in the rolling surface zone (b).*

Estimation of the size of the actual grain

The estimation of the actual grain size was determined from the "rough" surface of the non-working fillet along the border of the ferrite mesh of the decarburized layer. The distribution by grain numbers is shown in Fig. 7.13. From the presented distributions it can be seen that the main share of grains (71.4%) is estimated at No. 7-8 of GOST 5639-82. In this case, the average value of the grain number is 7.29, which corresponds to the average grain size of 29.8 μm (Table 7.2).

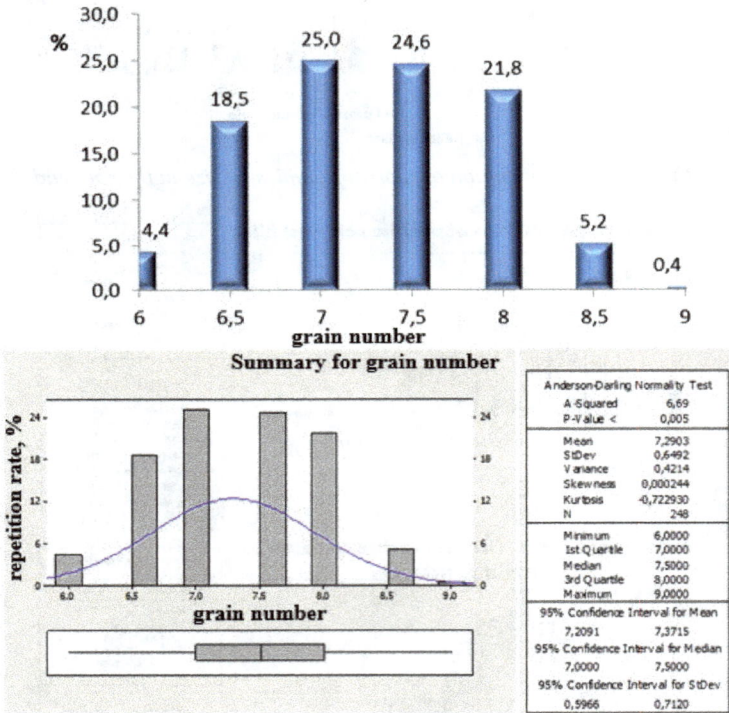

Figure 7.13 *Distribution of grains by size in the head.*

Thus, the studies carried out by the methods of optical and scanning electron microscopy at the macro- and microlevels of the structure of a fragment of a rail of the R350HT

Structure and Properties of Lengthy Rails after Extreme Long-Term Operation
Materials Research Foundations **106** (2021)

Materials Research Forum LLC
https://doi.org/10.21741/9781644901472

category removed from the track after the hauled load of 1411 million gross tons, showed that:

1. In terms of chemical composition, the metal of the investigated rail fragment meets the requirements of TU 0921-276-01124323-2012 for steel grade E76HF.

2. The microstructure of the rail fragment metal is represented by a finely dispersed lamellar pearlite of 2.0 points with inclusions of excess ferrite along the grain boundaries, estimated at 1.5 points of scale No. 7 GOST 8233. Bainite was not detected in the rail metal microstructure.

3. The value of the inter-plate distance in the railhead varies within (0.10-0.15) microns. The average size of pearlite colonies in the fillet zone is 6.2 microns, and on the rolling surface is 5.6 microns. The main array of actual grain size values, estimated only in the non-working fillet zone, was No. 7-8 of GOST 5639-82, which corresponds to an average grain size of 29.8 μm.

7.2 Metal structure of rails at submicro- and nanoscale

This section presents the results obtained in the study of the structure of the rail metal in a layer located at a distance of 10 mm from the rolling surface and fillet by scanning and transmission diffraction electron microscopy. The performed studies have shown that the main structural component of steel is the grains of highly dispersed pearlite of lamellar morphology (Fig. 7.14).

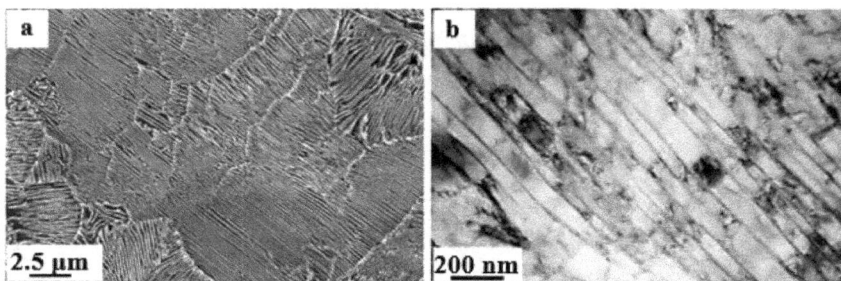

Figure 7.14 The structure of lamellar pearlite, revealed by the methods of scanning electron microscopy of the etched section (a), and by the methods of transmission electron microscopy of thin foils (b).

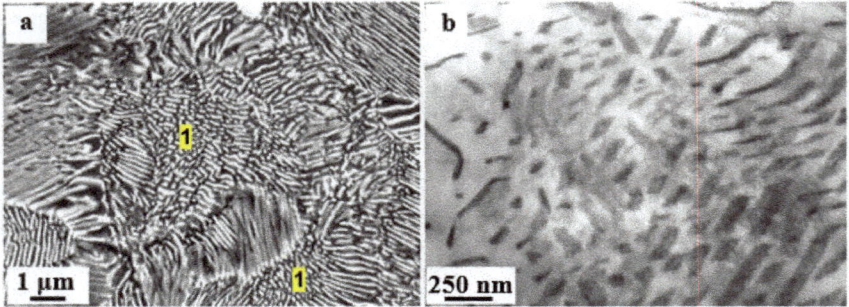

Figure 7.15 *The structure of degenerate pearlite, discovered by the methods of scanning electron microscopy of the etched section (a), and by the methods of transmission electron microscopy of thin foils (b). In (a) the grains of the degenerate pearlite are indicated by "1".*

The second morphological component of the steel structure is grains of degenerate pearlite, where the cementite has a globular shape, or the shape of short plates (Fig. 7.15).

The third structural component of the investigated steel is the grains of structurally free ferrite (i.e., ferrite grains that do not contain particles of the carbide phase) (Fig. 7.16). As a rule, grains of structurally free ferrite are arranged in the form of interlayers along the grain boundaries of pearlite.

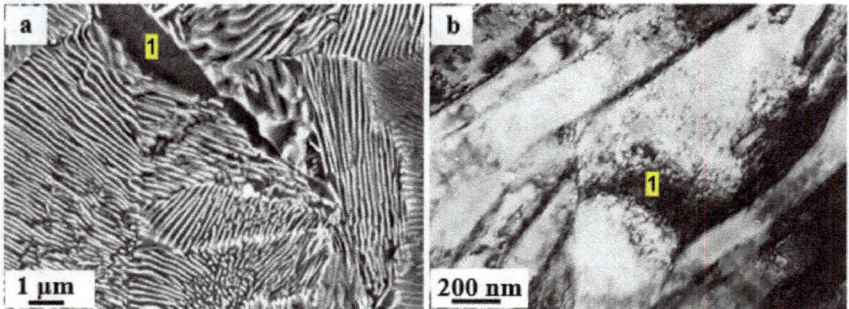

Figure 7.16 *Grains of structurally free ferrite, shown by scanning electron microscopy of etched sections (a), and by transmission electron microscopy of thin foils (b). Grains of structurally free ferrite are indicated by "1".*

Structure and Properties of Lengthy Rails after Extreme Long-Term Operation
Materials Research Foundations **106** (2021)

Materials Research Forum LLC
https://doi.org/10.21741/9781644901472

It was established by transmission electron microscopy that a dislocation substructure is present in the volume of grains of structurally free ferrite (Fig. 7.16, b) and the ferrite component of pearlite colonies (Fig. 7.17, a).

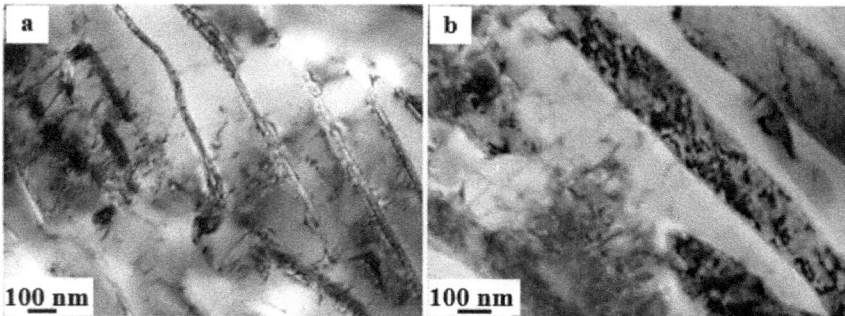

Figure 7.17 Electron-microscopic image of the dislocation substructure of a lamellar pearlite colony. Transmission electron microscopy of thin foils.

Dislocations are predominantly distributed chaotically, less often they form dislocation clusters. Cementite plates of pearlite colonies are also defective (Fig. 7.17, b). The contrast detected in the volume of cementite plates suggests the formation of misoriented regions (fragments) in them.

The metal of the rails is in an elastically stressed state. When studying the structure of steel by transmission electron microscopy of thin foils, this fact manifests itself in the bending-twisting of the crystal lattice [112, 121, 152, 154, 228], which is accompanied by the appearance of bending extinction contours on electron microscopic images (Fig. 7.18, the contours are indicated by arrows). As a rule, the contours intersect the pearlite colony, passing from one ferrite plate to another, which indicates the bending-twisting of the colony as a whole (Fig. 7.18, a). Much less often, the bending contour is located within one ferrite plate, which indicates bending-torsion of the crystal lattice of this plate (Fig. 7.18, b).

Structure and Properties of Lengthy Rails after Extreme Long-Term Operation
Materials Research Foundations **106** (2021)

Materials Research Forum LLC
https://doi.org/10.21741/9781644901472

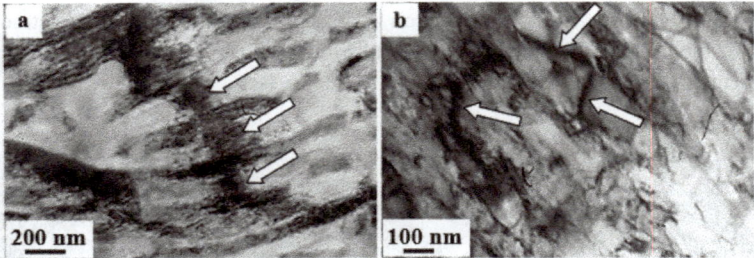

Figure 7.18 Electron microscopic image of the structure of lamellar pearlite with bending extinction contours.

7.3 Gradient of the rail metal structure after the hauled load of 1,411 million gross tons

The operation is accompanied by the formation of the structural-phase state of the inhomogeneous metal in the thickness of the head, i.e. the formation of a gradient structure. The structure gradient was studied using transmission electron diffraction microscopy. The objects of investigation for TEM were made by the methods of electrolytic thinning of plates located at the rolling and fillet surfaces and at a distance of 2 and 10 mm from the surface, cut out by the methods of electrospark erosion of the metal. The sample preparation scheme is shown in Figs. 2.1 and 7.19.

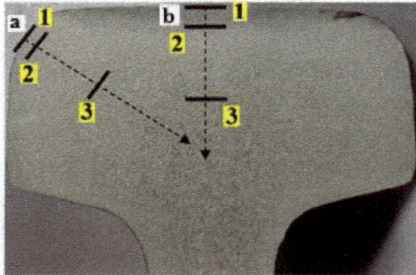

Figure 7.19 Scheme of preparation of a sample in the manufacture of objects of the steel by methods of transmission electron microscopy. Notations: 1 - rolling surface, 2 - layer at a distance of 2 mm from the surface, 3 - layer at a distance of 10 mm from the surface; (a) fillet; (b) central axis.

The gradient structure of the rail metal formed along the central axis of the head.

As shown in Figs. 7.5 and 7.6, single small discontinuities with a depth of up to 30 μm were found on non-etched thin sections prepared from the rolling surface of the head by optical microscopy. The depth of deformation from the rolling surface is insignificant and does not exceed 35 microns.

As a result of the studies performed by the methods of transmission electron microscopy, it was established that the steel layer at a depth of 10 mm is represented mainly by pearlite of lamellar morphology, the characteristic image of which is shown in Fig. 7.20 (a). Cementite plates in pearlite colonies are curved and separated by ferrite bridges [213]. The relative content of lamellar pearlite is 0.7 of the steel structure. A dislocation substructure is found in the ferrite component of pearlite colonies. The dislocations are located randomly or form dislocation clusters (Fig. 7.20, b). The scalar dislocation density determined by the randomly thrown secant method [113] is $<\rho> = 2.54 \times 10^{10}$ cm^{-2}.

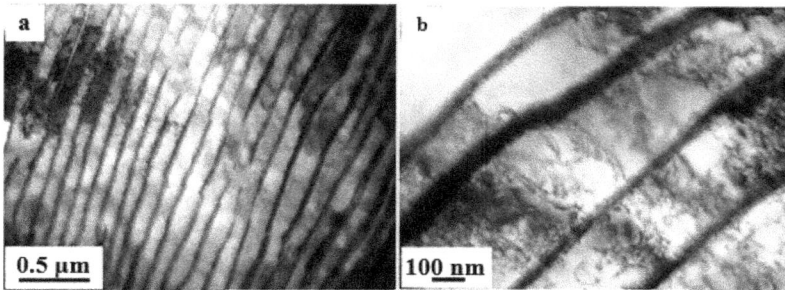

Figure 7.20: Metal structure in a layer located at a distance of 10 mm from the rolling surface.

Bending extinction contours are observed on electron microscopic images of pearlite grains, (Fig. 7.21, contours are indicated by arrows) [112].

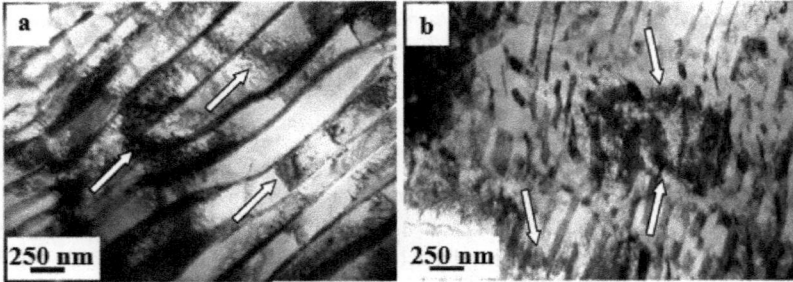

Figure 7.21: *Flexural extinction contours in a layer located at a distance of 10 mm from the rolling surface. The contours are indicated by arrows.*

The presence of such contours indicates the curvature-torsion of the crystal lattice of the material [229]. Techniques developed at Tomsk State University of Architecture and Civil Engineering (TSUACE) in a team led by E.V. Kozlov and N.A. Koneva, make it possible to estimate the excess density of dislocations leading to bending-twisting of the crystal lattice of the investigated foil region by using bending extinction contours [118, 151, 221]. The performed estimates show that the excess density of dislocations in a layer of rail steel located at a distance of 10 mm from the rolling surface is $\rho = 1.7 \times 10^{10}$ cm^{-2}.

In a significantly smaller amount (0.25), the grains of degenerate pearlite, in which the cementite has a globular shape (Fig. 7.22), and grains of structurally free ferrite (0.05 of the steel structure) are found in the structure of the studied steel layer. Two morphological components of degenerate pearlite are found. Firstly, grains in which cementite particles have a globular morphology (Fig. 7.22, a), and, secondly, grains in which cementite particles are in the form of short plates (Fig. 7.22, b).

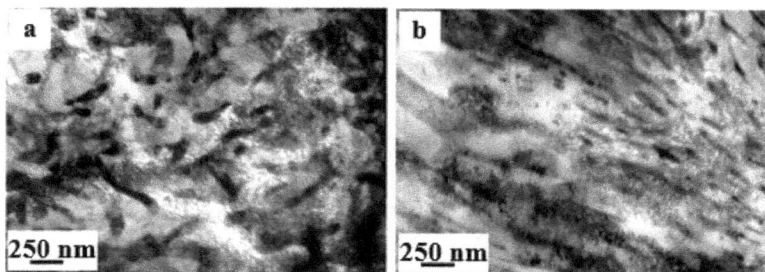

Figure 7.22 *Structure of degenerate metal pearlite in a layer located at a distance of 10 mm from the rolling surface.*

The structure of the layer located at a distance of 2 mm from the rolling surface is characterized by some features that distinguish it from the structure of the layer located at a depth of 10 mm. These features include, first, lamellar pearlite, in the volume of the colonies of which cementite plates are cut and displaced relative to each other (Fig. 7.23). The sizes of displaced areas of cementite plates vary from 50 to 500 nm. The size of the fragments of cementite plates will depend on the degree of involvement of this grain in the deformation processes that take place during the operation of rails.

Second, the formation of a fragmented (subgrain) structure in grains of degenerate pearlite (Fig. 7.24). The presence in the microelectron diffraction patterns (Fig. 7.24, b) of the strands on the reflections of the α-phase, obtained from such a structure, makes it possible to estimate the azimuthal component of the angle of complete disorientation of the structure, reaching five degrees.

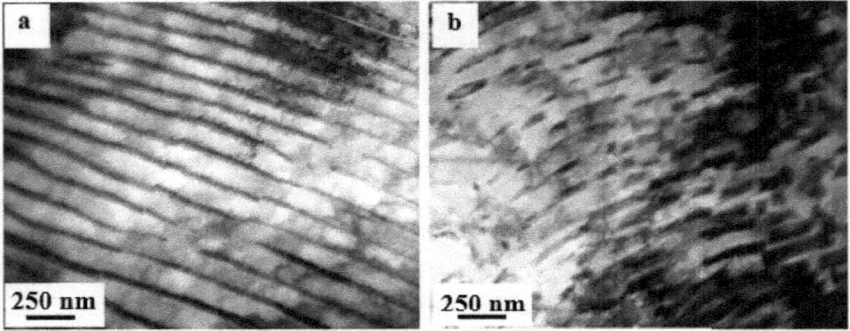

Figure 7.23 *Electron microscopic image of the structure of pearlite lamellar morphology of the layer located at a depth of 2 mm.*

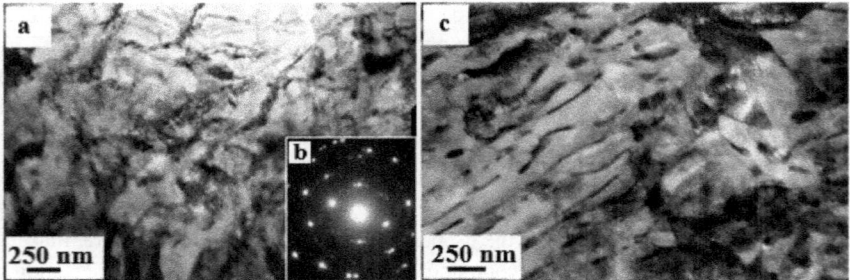

Figure 7.24 *Electron microscopic image of the subgrain (fragmented) structure in the grains of a degenerate pearlite layer located at a depth of 2 mm.*

Third, the destruction of cementite plates of lamellar pearlite by dissolving them with the escape of carbon atoms to dislocations, followed by precipitation in the form of nanosized particles of the carbide phase in the volume of ferrite plates (Fig. 7.25). Particle sizes range from 8 to 12 nm. Particles of cementite of a rounded, less often lamellar shape are usually located on dislocation lines, fixing them and forming a reticular dislocation substructure (Fig. 7.25). In some cases, the formation of regions in the pearlite structure containing nanoscale particles is discovered (Fig. 7.26). It is apparent that the above options for transforming the steel structure are a consequence of the deep plastic deformation of the material.

Structure and Properties of Lengthy Rails after Extreme Long-Term Operation
Materials Research Foundations **106** (2021)

Materials Research Forum LLC
https://doi.org/10.21741/9781644901472

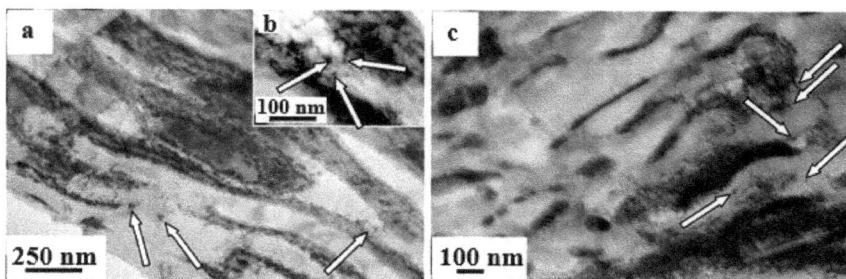

Figure 7.25 *Electron microscopic image of nanosized particles (particles are indicated by arrows) in the structure of pearlite with lamellar morphology of a layer located at a depth of 2 mm.*

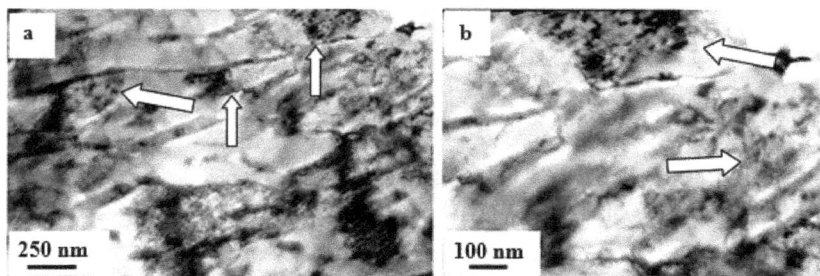

Figure 7.26 *Electron microscopic image of fragments containing nanosized particles (fragments are indicated by arrows) in the structure of pearlite with lamellar morphology of a layer located at a depth of 2 mm.*

A typical electron microscopic image of the structure of the metal layer forming the rolling surface is shown in Fig. 7.27. It is clearly seen that the result of the long-term operation is a significant transformation of the pearlite structure. Along with the colonies that retained the morphology of lamellar pearlite, a subgrain-type structure is formed in the surface layer, the relative content of which is 0.55 of the steel structure (Fig. 7.27, a). The sizes of subgrains vary from 100 to 150 nm. Along the boundaries of subgrains and at the junctions of the boundaries, there are particles of the carbide phase, the sizes of which vary from 30 to 55 nm. Quite often, particles of the second phase are located in the volume of subgrains on dislocation lines (Fig. 7.27, b). The sizes of such particles vary from 10 to 15 nm.

Structure and Properties of Lengthy Rails after Extreme Long-Term Operation
Materials Research Foundations **106** (2021)

Materials Research Forum LLC
https://doi.org/10.21741/9781644901472

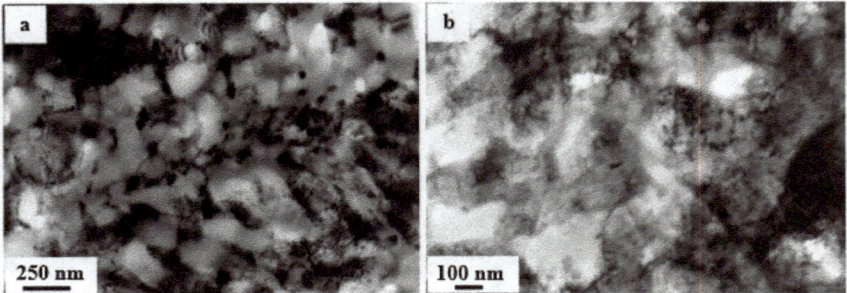

Figure 7.27 *Electron microscopic image of the structure of the layer adjacent to the rolling surface.*

The scalar density of dislocations in the structure of pearlite colonies $<\rho> = 3.7 \times 10^{10}$ cm^{-2}; in the subgrain structure $<\rho> = 3.0 \times 10^{10}$ cm^{-2}. It can be noted when comparing with the above results, that the value of the scalar dislocation density of the surface layer is 1.5 times higher than the scalar dislocation density of the layer located at a depth of 10 mm. In the same way (1.5 times) the excess density of dislocations in the layer that forms the rolling surface also increases.

The structural-phase state of the rolling surface metal was studied by X-ray phase analysis. Typical X-ray patterns of the material under study are shown in Fig. 7.28.

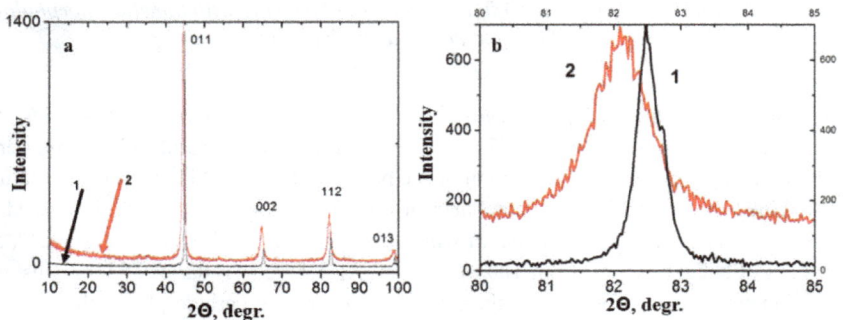

Figure 7.28 *Sections of X-ray patterns obtained from the volume of steel located at a distance of 20 mm from the rolling surface ((1), hereinafter referred to as the initial state), and from the rolling surface (2). The indices of the diffraction lines of the α-phase are shown in (a); (b) shows the diffraction maxima of [112] α-Fe.*

X-ray phase analysis shows that the main phases of the investigated steel are a solid solution based on α-iron (body-centered cubic crystal lattice) and iron carbide (Fe_3C, cementite, orthorhombic crystal lattice). The crystal lattice parameter of α-Fe in the initial state is $a_{init} = 0.28693$ nm; lattice parameter of α-Fe metal of rolling surface $a_{rs} = 0.28699$ nm. An increase in the crystal lattice parameter of the surface layer of the α-phase may be associated with an increase in the concentration of carbon atoms in the solid solution. Simultaneously with an increase in the crystal lattice parameter, a significant broadening of the diffraction lines is observed (Fig. 7.28, b). The broadening of diffraction lines can be caused by the small size of scattering crystallites and significant microdistortions (stresses of the second kind) in crystallites. Therefore, analysis of the shape of diffraction peaks, their displacements, and broadenings is used to determine the average sizes of coherent scattering regions (CSRs) and the magnitude of microstrains [230]. However, it should be taken into account that if the CSR size (D) is large and/or the value of microdistortions of the crystal lattice (ε) is small, then it is impossible to calculate these characteristics of the material. Therefore, there is a minimum value of D and a maximum value of ε, which can be determined from line broadenings: 0.005 μm < D < 0.2 μm and $10^{-4} < \varepsilon < 10^{-2}$. The analysis of the diffraction patterns performed in this work showed that for the steel of the initial state it is not possible to determine the values of the sizes of the coherent scattering regions D_{init} and the magnitude of microdistortions $(\Delta d/d)_{init}$. This may mean that $D_{init} > 0.2$ μm, and $(\Delta d/d)_{init} < 10^{-4}$. For the rolling surface, these values were determined and are D = 22.06 nm and $\Delta d / d = 1.562 \times 10^{-3}$. Thus, after the hauled load of 1411 million gross tons, the metal of the rolling surface of the rails is characterized by a relatively small value of the coherent scattering regions and relatively large values of the microdistortions of the crystal lattice of the α-phase.

Thus, the long-term operation is accompanied by a significant transformation of the metal structure of the railhead along the central axis. The analysis of transverse thin sections revealed single small discontinuities up to 30 microns deep. It is shown that the depth of deformation from the rolling surface, detected by metallography methods of the etched section, is insignificant and does not exceed 35 μm. It has been established that the long-term operation is accompanied by the formation of a gradient structure. At the submicro- and nanoscale, this process leads to the following changes in the structural-phase state of the material. First, to the destruction of the lamellar pearlite structure and the formation of subgrain structure of submicron (100-150 nm) sizes in the volume of pearlite colonies. Secondly, to precipitation along the boundaries and in the volume of subgrains of particles of the carbide phase of the nanometer range. Third, to the growth of microdistortions and the crystal lattice parameter of the solid solution based on α-iron.

Fourth, to strain hardening of the metal, leading to an increase (1.5 times) relative to the initial state, scalar, and excess dislocation density.

Gradient metal structure formed along the axis of symmetry of the working fillet of the railhead.

As already noted, the structure of the metal in the layer located at a distance of 10 mm from the fillet surface is formed by pearlite grains of lamellar morphology, the relative content of which is 0.79 (Fig. 7.29, a). In a smaller amount (0.18 metal structure) there are regions of "degenerate pearlite" (Fig. 7.29, b), the rest are grains of structurally free ferrite (ferrite grains, in the volume of which there are no cementite particles).

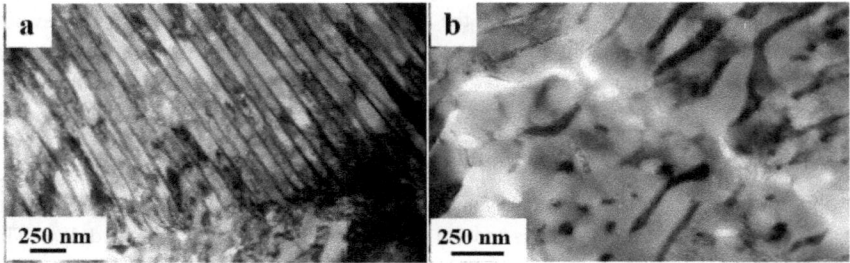

Figure 7.29 *Metal structure in a layer located at a distance of 10 mm from the head surface along the axis of symmetry of the working fillet.*

At a distance of 2.0 mm (and very rarely, at a distance of 10 mm) from the surface of the working fillet, deformed pearlite containing plates of cementite, destroyed into separate particles displaced relative to each other, is added to the above structural components of steel (Fig. 7.30).

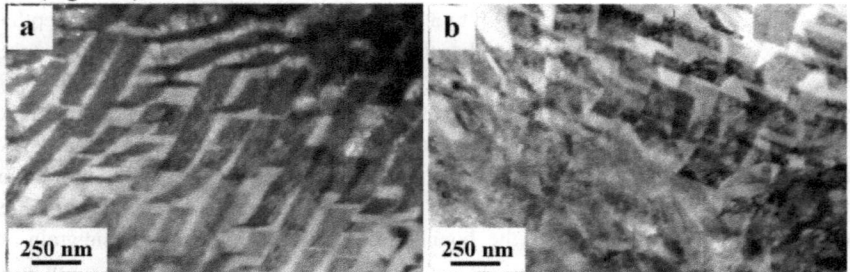

Figure 7.30 *TEM image of a metal structure (pearlite lamellar morphology, destroyed) in a layer at a depth of 2 mm.*

The second mechanism of destruction of cementite plates, as already discussed above, is their dissolution with the escape of carbon atoms to dislocations, followed by precipitation in the form of nanosized particles of the carbide phase in the volume of ferrite plates (Fig. 7.31). The sizes of the formed particles vary from 10 to 15 nm. Particles of cementite of a rounded, less often lamellar shape are usually located on dislocation lines, fixing them and forming a reticular dislocation substructure (Fig. 7.31).

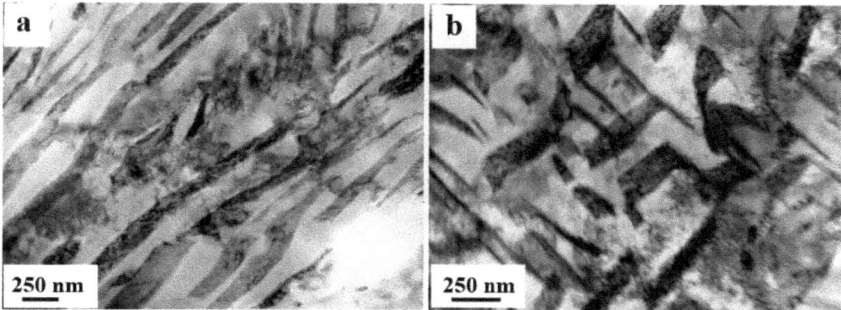

Figure 7.31 TEM image of the structure of pearlite of lamellar morphology, which is destroyed by the dissolution of cementite plates, followed by the release of nanosized particles of the carbide phase; layer located at a distance of 2 mm from the surface of the working fillet.

In the surface layer of the fillet, in addition to the above, a structure is formed, which we call a "ferrite-carbide mixture" (Fig. 7.32). A characteristic feature of this structure is the nanoscale range of grains, subgrains, and particles of the carbide phase forming it (Fig. 7.33). The size of the grains and subgrains forming this type of structure varies within 40 - 70 nm (Fig. 7.33, a). The size of the particles of the carbide phase located along the grain and subgrain boundaries varies within 8 - 20 nm (Fig. 7.33, b).

Figure 7.32 *The structure of a ferrite-carbide mixture (surface layer of the metal of the working fillet).*

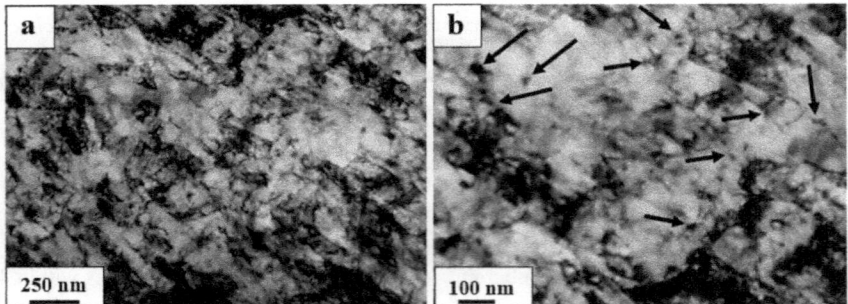

Figure 7.33 *TEM image of the "ferrite-carbide mixture" structure forming in the surface layer of the working fillet. Arrows in (b) indicate particles of the carbide phase.*

Thus, studies of the rail metal structure carried out along the axis of symmetry of the working fillet show that the operation of the rails is accompanied by the formation of a gradient of structural components, which consists in a regular decrease as the fillet surface is approached. Furthermore, the relative volume content of the material with the lamellar pearlite structure and an increase in the relative content volume of material with the structure of destroyed pearlite and the ferrite-carbide mixture is formed.

Conclusion

Investigations at the macro and micro levels, carried out by the methods of modern physical materials science, have shown that, in terms of chemical composition, the metal of the investigated rail fragment meets the requirements of TU 0921-276-01124323-2012 for steel grade E76HF. The metal microstructure of the rail fragment is represented by finely dispersed lamellar pearlite of 2.0 points with inclusions of excess ferrite along the grain boundaries, estimated at 1.5 points of scale No. 7 of GOST 8233. Bainite was not detected in the microstructure of the rail metal. The value of the inter-plate distance in the railhead varies within 0.10-0.15 microns. The average size of pearlite colonies in the fillet zone is 6.2 microns and on the rolling surface 5.6 microns. The main array of actual grain size values, estimated only in the non-working fillet zone, was 7-8 number GOST 5639-82, which corresponds to an average grain size of 29.8 μm.

It has been established [231-236] that the long-term operation of rails is accompanied by the formation of a gradient structure, which is expressed in a significant transformation of the metal structure of the head along the central axis and along the axis of symmetry of the working fillet. Single small discontinuities with a depth of up to 30 μm were detected in the study of the metal structure along the axis of symmetry and the analysis of transverse thin sections. It is shown that the depth of deformation from the rolling surface, detected by the metallography of etched sections, is insignificant and does not exceed 35 μm. At the submicro- and nanoscale, extremely long-term operation leads to the following changes in the structural-phase state of the material. First, to the destruction of the lamellar pearlite structure and the formation of subgrain structure of submicron (100-150 nm) sizes in the volume of pearlite colonies. Secondly, to precipitation along the boundaries and in the volume of subgrains of particles of the carbide phase of the nanometer range. Third, to the growth of microdistortions and the crystal lattice parameter of the solid solution based on α-iron. Fourth, to strain hardening of the metal, leading to an increase (1.5 times) relative to the initial state, scalar, and excess dislocation density. Studies of the gradient of the rail metal structure, carried out along the axis of symmetry of the working fillet, show that the operation of the rails is accompanied by a natural decrease as the fillet surface is approached, by the relative volume content of the material with the lamellar pearlite structure and by an increase in the relative volume content of the material with the structure of destroyed pearlite and ferrite-carbide mixture.

Structure and Properties of Lengthy Rails after Extreme Long-Term Operation
Materials Research Foundations **106** (2021)

Materials Research Forum LLC
https://doi.org/10.21741/9781644901472

8. Physical Nature of Strengthening of Rail Metal after Extremely Long Operation

The mechanical properties of the rail metal were characterized by impact strength, hardness, and microhardness. The impact strength was determined at a test temperature of +20°C on two specimens of type one according to GOST9454. The specimens were cut from the head. The hardness was measured on the rolling surface and along the head section following the requirements of TU 0921-276-01124323-2012. Additionally, the hardness was measured in the upper part of the neck (about 30 mm above point 6 of the requirements of clause 1.8.1 of TU 0921-276-01124323-2012).

8.1 Impact strength and hardness

The results of testing the metal for impact strength and hardness are shown in Table 8.1.

Table 8.1 *Results of mechanical tests.*

Smelting number	KCU +20°C J / cm²		Hardness, HB					
			RS	10 mm	Fillet		22 mm	Neck
					No.1	No.2		
29861	30	27	388-399	381	364	362	373	345
TU 0921-276-01124323-2012 requirements for R350HT category rails	Not less 15		363-401	Not less 341				Not more 341

From the test results presented in Table 8.1, it can be seen that in terms of impact strength and hardness on the rolling surface of the head and along its cross-section, the metal of the test sample meets the requirements of TU 0921-276-01124323-2012 for rails of the R350HT category. One of the hardness values measured on the rolling surface of the head at a distance of about 15 mm from the rail end has a higher number, which is probably due to work-hardening processes with an increased impact of the wheel on the rail in the indicated zone. The hardness measured in the neck has slightly increased values relative to the requirements of the technical specifications.

Additionally, the hardness of the metal of the sample was measured along the cross-section of the head in the transverse direction by the Rockwell method at a distance of 2, 10, and 22 mm from the rolling surface of the head along the vertical axis of symmetry and from the fillets. The results of determining the hardness are presented in Table 8.2.

Table 8.2 *The results of determining the hardness of the metal, performed by the Rockwell method.*

Smelting number	Measurement location	Hardness HRC, at a distance from the surface, mm		
		2	10	22
29861	Working fillet	38.5	35.5	34.8
	Center	37.1	35.8	35.6
	Non-working fillet	35.3	35.5	35.2

From the analysis of the results presented in Table. 8.2, it follows that the hardness values in the central zone and the working fillet at a depth of 2 mm are at a higher level (38.5 - 37.1 HRC) compared to the non-working fillet (35.3 HRC), which is due to the presence of deep deformation in the indicated zone, accompanied by metal melting. At a depth of 10 and 22 mm from the rolling surface, the hardness of the metal is characterized by a lower level (by 2-3 HRC) compared to the depth of 2 mm and has comparable values (34.8 - 35.8 HRC).

8.2 Microhardness

Microhardness was determined with a PMT-3 device by the Vickers method with a load on the indenter of 0.5 N along the vertical axis of symmetry of the tread and the fillet surface in the interval 10 to 110 μm with a step of 10 μm (two tracks), at a depth of 2 and 10 mm from the surface at the location of both fillets and the central zone of the rolling surface of the sample head according to the results of four measurements in each zone. The averaged values obtained during the measurement are presented in Table. 8.3.

From the analysis of the test results presented in Table 8.3, it follows that the microhardness at a depth of 2 mm, determined along the central axis and the axis of symmetry of the fillet, has similar values (1481 - 1486 MPa). Further, to a depth of 10 mm, the microhardness value decreases to 1210 – 1385 MPa. One of the reasons for this is an increase in the interlamellar distance (decrease in dispersion) of pearlite colonies.

The microhardness profiles shown in Fig. 8.1 indicate that the operation of rails is accompanied by a significant strengthening of the near-surface layer of steel up to 80 - 100 microns thick, regardless of the investigated section of the rail (rolling surface or working fillet). At the same time, near the rail surface, the hardness of steel is 1.5 - 2 times higher relative to the volume of the product and decreases with distance from the working surface.

Table 8.3 *Values of microhardness of the head metal, determined at a distance of 2 and 10 mm from the surface at the place of both fillets and the central zone of the rolling surface of the sample head.*

Measurement area	Microhardness, MPa, at a distance from the surface	
	2 mm	10 mm
Working fillet	1475	1385
Non-working fillet	1486	1274
Rolling surface	1481	1210

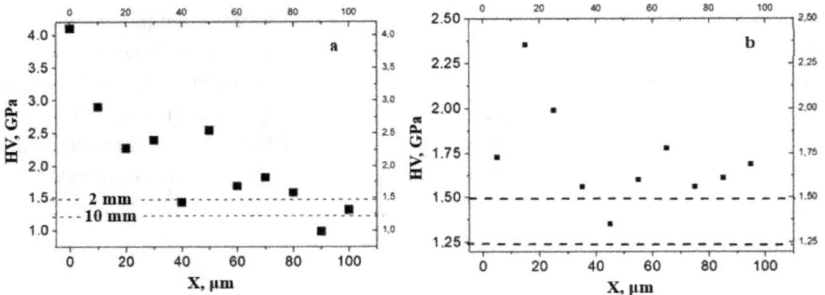

Figure 8.1 *Dependence of microhardness on the distance from the rolling surface (a) and the fillet surface (b), obtained along the vertical axis of symmetry of the rolling surface and the fillet surface under a load on the indenter of 0.5 N. The microhardness of steel at a distance of 2 mm is 1.48 GPa, 10 mm - 1.21 GPa.*

8.3   Tribological tests of steel

As mentioned in Chapter 2, tribological studies (determination of the wear parameter and the coefficient of friction) were carried out on a Pin on Disc and Oscillating TRIBOtester (TRIBOtechnic, France) with the following parameters: a ball made of ShX15 steel with a diameter of 6 mm, a track radius of 4 mm, a load on indenter 12 N, sample rotation speed 25.0 mm/s, at room temperature. The degree of wear of the material was determined from the results of profilometry of the track formed during the tests.

The test results presented in Table. 8.4 indicate that the wear resistance of the rolling surface has increased by approximately 1.7 times to the volume of the product.

Table 8.4 *Tribological characteristics of steel.*

Test area	μ	k, 10^{-5}, mm^3/Nm
Layer at a depth of 15 mm	0.6	2.0
Rolling surface	0.76	0.3

An increase in the wear resistance of steel is accompanied by an increase, by a factor of 1.3, in the friction coefficient (Table 8.4). The results of tribological tests shown in Fig. 8.2 indicate that the nature of the dependence of the friction coefficient on the test time is different for the volume of steel located at a depth of 15 mm from the rolling surface (Fig. 8.2, a), and the volume of steel forming the rolling surface (Fig. 8.2, b).

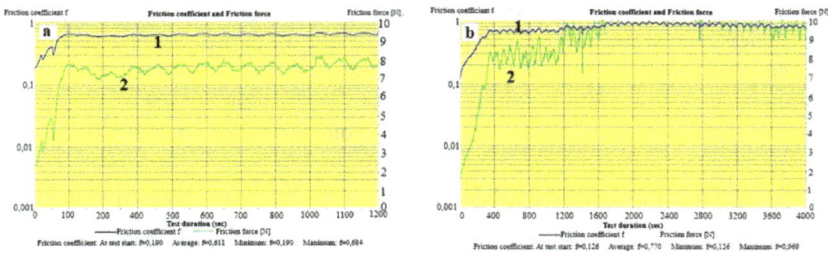

Figure 8.2 *Dependence of the friction coefficient (curve 1) and the friction force N (curve 2) on the test time; (a) tests were carried out on a cross-section of the rail at a distance of 15 mm from the rolling surface, (b) tests were carried out on the rolling surface.*

In the first case, the change in the friction coefficient reaches a stationary level after 100 seconds of running; in the second case, after 400 seconds. The latter indicates a change in the structural-phase state of steel in the surface layer during the operation. This is also indicated by the profiles of the friction tracks shown in Fig. 8.3. It is seen that the friction track obtained during tribological tests of the rail rolling surface has a smoother profile, which indicates a more equal strength state of the surface layer of the friction track in comparison with the material of the rail volume.

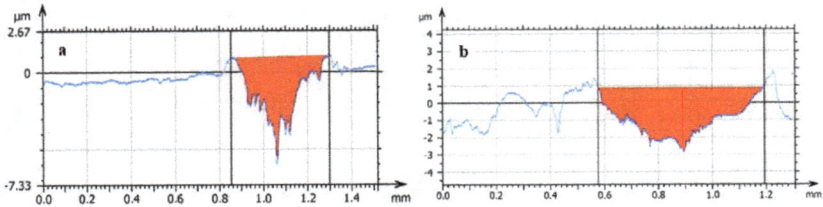

Figure 8.3 *Typical images of the profile of the friction track of rail steel at a distance of 10 mm from the rolling surface (a); the profile of the friction track formed during tribological tests of the rolling surface (b).*

Thus, as a result of the performed studies, it was found that the operation of rails is accompanied by a significant (1.5-2 times) strengthening of the near-surface layer of steel with a thickness of up to 80 - 100 microns, regardless of the investigated section of the rail (rolling surface or working fillet). An increase in the wear resistance of the rolling surface by approximately 1.7 times relative to the volume of the product is discovered.

8.4 Physical nature of metal hardening during extremely long-term operation

Studies of rail metal, the results of which were discussed in detail in Chapters 3, 4, and 6 showed that the structure of the initial steel is represented by grains of pearlite of plate morphology, grains of degenerate pearlite (ferrite-carbide mixture), and, in a very small amount, grains of structurally free ferrite (ferrite grains not containing cementite particles).

The long-term operation is accompanied by the formation of a gradient structure. At the submicro- and nanoscale, this process leads to the following changes in the structural-phase state of the material. First, to the destruction of the lamellar pearlite structure and the formation of subgrain structure of submicron (100-150 nm) sizes in the volume of pearlite colonies. Secondly, to precipitation along the boundaries and in the volume of subgrains of particles of the carbide phase of the nanometer range. Thirdly, to the growth of microdistortions and the crystal lattice parameter of a solid solution based on α-iron. Fourth, to strain hardening of the metal, which leads to an increase (by 1.5 times) relative to the initial state, scalar, and excess dislocation density. Studies of the gradient of the rail metal structure, carried out along the symmetry axis of the working fillet and along the central axis, show that the operation of the rail is accompanied by a natural decrease of the relative volume content of the material with the lamellar pearlite structure as the fillet surface is approached, and the increase in the relative volume content of the material with the destroyed structure of perlite and ferrite-carbide mixture.

The discovered transformations of the steel structure will have a significant effect on the strength and plastic characteristics of the metal, ultimately determining the service life of the product. Evaluation of hardening mechanisms can help us to obtain the regularities connecting the parameters of the structure and the strength properties of the material, and to reveal the physical nature of the process of evolution of the properties. Estimates of the magnitude of the hardening mechanisms were carried out using widely tested expressions considered earlier in several works [237-255]. These expressions correspond to relations 6.1 - 6.9 (Chapter 6).

As noted above, the main structural component of steel is pearlite grains of lamellar morphology, a typical image of which is shown in Fig. 8.4.

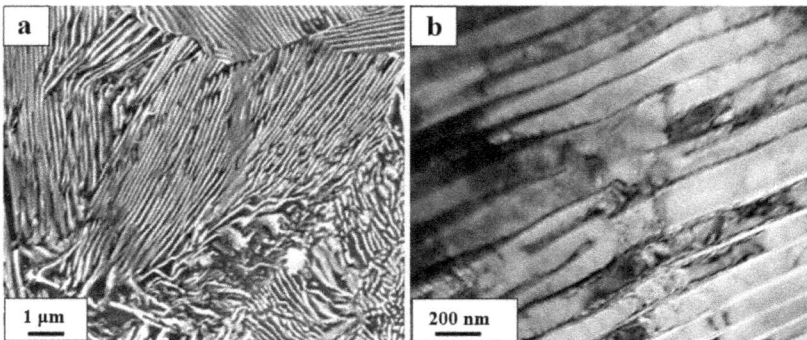

Figure 8.4 Pearlite structure of metal: (a) scanning electron microscopy; (b) transmission electron microscopy.

The value of the yield stress of pearlitic steel can be estimated by the expression (6.7) [187, 256]. The relative perlite content of the lamellar morphology of the rail metal after the hauled load of 1400 million gross tons is shown in Fig. 8.5.

Structure and Properties of Lengthy Rails after Extreme Long-Term Operation
Materials Research Foundations **106** (2021)

Materials Research Forum LLC
https://doi.org/10.21741/9781644901472

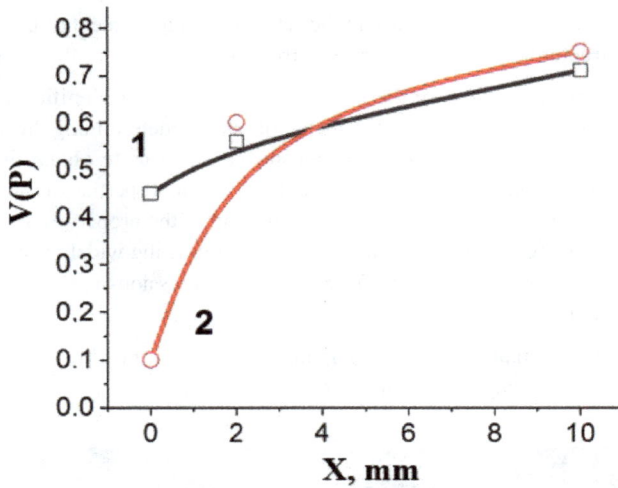

Figure 8.5 *Dependence of the relative content of pearlite lamellar morphology on the distance from the railhead surface; (curve 1) along the central axis; (curve 2) along the axis of symmetry of the fillet.*

It was found by transmission electron microscopy that a dislocation substructure is present in the ferrite component of pearlite colonies (Fig. 8.6). The dependence of the scalar dislocation density on the distance from the railhead surface is shown in Fig. 8.7.

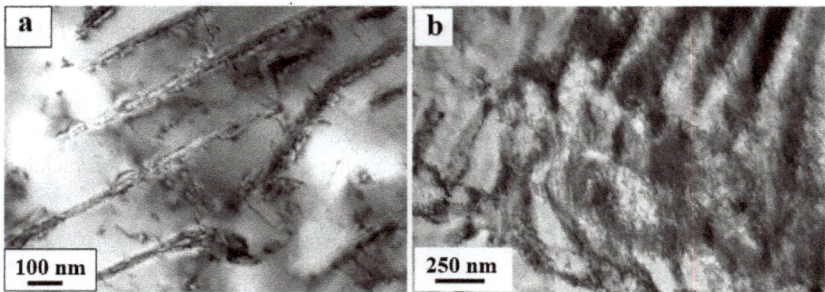

Figure 8.6 *Dislocation metal substructure observed in the layer at a depth of 2 mm from the surface of the working fillet; (a) lamellar perlite; (b) degenerate perlite.*

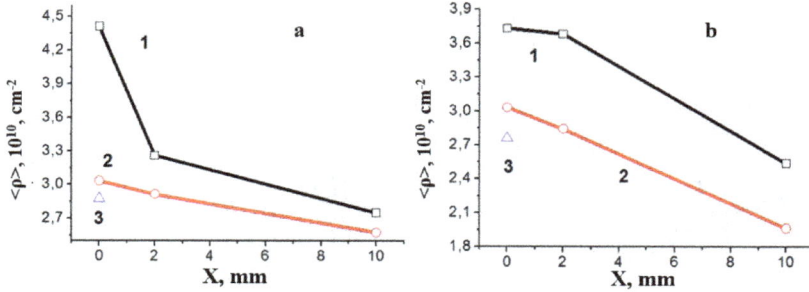

Figure 8.7 *Dependences of the scalar dislocation density on the distance from the head surface; (a) along the axis of symmetry of the working fillet; (b) along the central axis; (lines) 1: dislocation structure of lamellar pearlite; 2: degenerate pearlite; 3: ferrite-carbide mixture.*

The stress required to maintain plastic deformation, i.e. the flow stress σ required for moving dislocations (deformation carriers) to overcome the forces of interaction with stationary dislocations ("forest" dislocations) is related to the scalar dislocation density as in (6.2) [187, 224, 256].

The operation of rails is accompanied by the formation of internal stress fields in the steel. When studying the structure of steel using transmission electron microscopy, the presence of stress fields in the material, as noted earlier, manifests itself in the appearance of bending extinction contours on the electron microscopic images, indicating the curvature-torsion of the crystal lattice of this foil region (Fig. 8.8).

The procedure for assessing the magnitude of internal stress fields is reduced to determining the gradient of curvature-torsion of the crystal lattice χ, given in (6.3) [14, 152-154, 221].

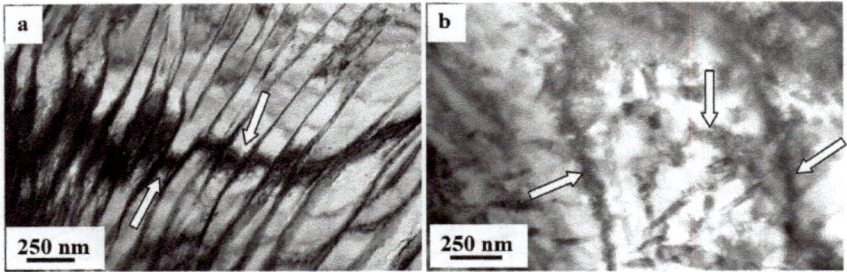

Figure 8.8 *Electron microscopic image of the structure of lamellar pearlite (a) and degenerate pearlite (b); the arrows indicate the bending extinction contours.*

The value of the excess dislocation density $\rho_\pm$ is related to χ through the Burgers vector of dislocations b based on the relation (6.4).

The dependence of the excess dislocation density on the distance from the head surface is shown in Fig. 8.9.

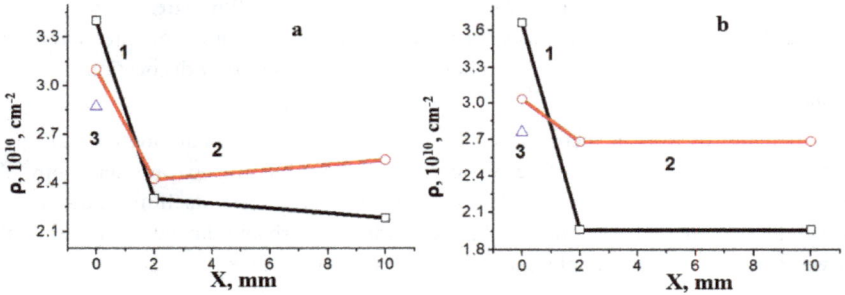

Figure 8.9 *Dependences of the excess dislocation density on the distance from the head surface; (a) along the axis of symmetry of the working fillet; (b) along the central axis; (lines) 1: dislocation structure of lamellar pearlite; 2: degenerate pearlite; 3: ferrite-carbide mixture.*

The value of the plastic component of internal stress fields is estimated based on the ratio (6.5) [14, 152-154, 221].

The operation of rails is accompanied by the process of dynamic aging of steel, which leads to the formation of nanosized particles of iron carbide in the material of a predominantly globular force (Fig. 8.10). Besides, globular particles of the carbide phase are also present in the grains of degenerate pearlite (Fig. 8.10, b).

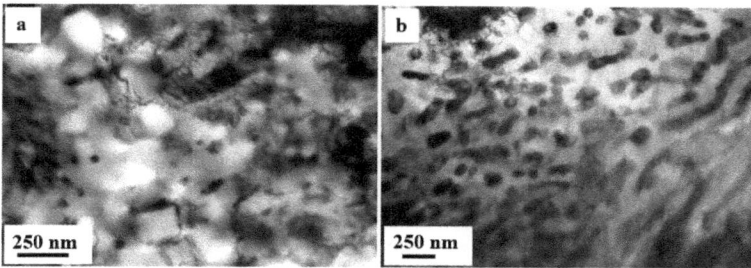

Figure 8.10 Electron microscopic image of the rail metal structure, demonstrating the presence of globular particles in the grains of the ferrite-carbide mixture (a) and the grains of degenerate pearlite (b).

Iron carbide particles larger than 5 nm lose their coherent connection with the crystal lattice of the α-phase [214, 256]. Consequently, the nanosized particles of the carbide phase present in the rail steel, the sizes of which exceed 10 nm, as well as the particles of degenerate pearlite, are incoherent. Incoherent particles, including cementite particles, are an obstacle to the movement of dislocations, which leads to the hardening of the material. Steel hardening estimates, taking into account the presence of incoherent particles of the second phase, are carried out using the relation (6.8) [226].

It was shown in Chapter 7 that the operation of rails is accompanied by the formation of a fragmented substructure in the surface layer. Typical structures formed in this case are shown in Fig. 8.11.

Figure 8.11 Electron microscopic image of the subgrain structure in the surface layer of the head.

The hardening of the material by low-angle boundaries (substructural hardening, hardening by the boundaries of fragments and subgrains) separating fragments (subgrains) can be estimated using the expression (6.1) [187, 213, 257].

By the methods of X-ray phase analysis, it was shown in Chapter 7 that operation is accompanied by an increase in the concentration of carbon in the crystal lattice of α-iron. It was found that the crystal lattice parameter of α-Fe in the initial state is a_{init} = 0.28693 nm; the lattice parameter of α-Fe rolling surface a_{rs} = 0.28699 nm. The relative content of cementite in the steel of the initial state is 3.39 wt.%; in the volume of steel forming the rolling surface is 3.31 wt.%.

Assuming that an increase in the crystal lattice parameter of α-iron is associated with the dissolution of cementite and the escape of carbon atoms into the solid solution, we estimate the carbon concentration in the solid solution based on α-Fe using the ratios proposed in [173]. The performed estimates show that the revealed increase in the crystal lattice parameter of α-iron can correspond to the transition to a solid solution of 0.0015 wt% carbon. Using the Fe-C phase diagram [173], it is possible to estimate the amount of released carbon corresponding to the obtained change in the relative content of cementite in steel. The performed estimates show that, in our case, the released carbon concentration is 0.0052 wt%. Using the results of estimated calculations, we can conclude that the main amount of carbon atoms released in the course of the destruction of cementite during the operation of the rail is deposited on crystal lattice defects (dislocations, interphase, and intraphase boundaries). Thus, based on the performed studies and estimated calculations, it can be concluded that the thermal deformation effect that occurs during the operation is accompanied by the destruction of cementite of the surface layer of steel, the release of carbon atoms, which is partially embedded in the crystal lattice of α-iron (insertion positions) and is deposited on defects crystalline structure of steel. Both processes undoubtedly lead to steel hardening [187, 153, 213, 223, 256].

The assessment of the solid solution hardening of steel due to carbon atoms and other alloying elements was carried out using an empirical expression of the form [215, 256]:

$$\sigma(ss) = \sum_{i=1}^{m} (k_i C_i) \tag{8.1}$$

where k_i is the coefficient of hardening of ferrite, which is the increase in the strength of the material at the yield point when dissolved in it 1 wt. % of the alloying element, C_i is the concentration of the elements dissolved in the ferrite, wt. %. The value of the coefficient k_i for various elements is determined empirically [203, 258].

The general yield stress of steel in the first approximation based on the additivity principle, which assumes the independent action of each of the material hardening mechanisms, can be represented as a linear sum of the contributions of individual hardening mechanisms [152, 153, 187, 215, 216, 256]:

$$\sigma = \Delta\sigma_0 + \Delta\sigma(L) + \Delta\sigma(\rho) + \Delta\sigma(h) + \Delta\sigma(cp) + \Delta\sigma(ss) + \Delta\sigma(P) \qquad (8.2)$$

where $\sigma_0 = 30$ MPa is the contribution due to friction of the matrix lattice, $\Delta\sigma(L)$ is the contribution due to the intraphase boundaries, $\Delta\sigma$ (ρ) is the contribution due to the dislocation substructure, $\Delta\sigma(cp)$ is the contribution due to the presence of particles carbide phases, $\Delta\sigma(h)$ is the contribution due to internal stress fields, $\Delta\sigma(ss)$ is the contribution due to solid solution strengthening, $\Delta\sigma(P)$ is the contribution due to the pearlite component of the steel structure.

Thus, having determined the quantitative characteristics of the steel structure, it is possible, as a first approximation, to analyze the physical mechanisms responsible for the evolution of the steel strength at the yield point during the operation of the rail, and also to identify the physical mechanisms of the formation of the strength gradient of the rail steel.

Using the results of the quantitative analysis of the steel structure given in Chapters 7 and 8, evaluations of the steel hardening mechanisms were carried out. The results of the evaluations are presented in Table. 8.5.

Analyzing the results shown in Table. 8.5, the following can be noted. First, the strength of steel is a multifactorial value and is determined by the combined action of several physical mechanisms. Secondly, the strength of the metal rail depends on the distance to the surface of the railhead, regardless of the place of analysis (along the central axis or the axis of symmetry of the fillet). Thirdly, the strength of the metal increases as it approaches the head surface; in this case, the strength of the metal along the axis of symmetry of the fillet is higher than along the central axis. Fourth, the main mechanism of metal hardening in the surface layer (in the layer forming the head surface) is the substructural one, caused by the interaction of moving dislocations with low-angle boundaries of fragments and nanosized subgrains.

Table 8.5 *Estimates of the mechanisms of strengthening of the structure formed at different distances along the central axis and the axis of symmetry of the fillet (hauled load of 1411 million gross tons). The contributions are given taking into account the volume fraction of the structure with this hardening mechanism.*

Parameters, average by material, MPa	Rolling surface			Working fillet		
	Distance from the surface, mm					
	10	2	0	10	2	0
$\Delta\sigma(P)$	142.5	161.5	85.5	152	152	95
$\Delta\sigma(L)$	0	0	473.3	0	0	1455.6
$\Delta\sigma(\rho)$	152.8	181	181.4	164	206	190.4
$\Delta\sigma(h)$	131.3	149	255	148.6	149.6	230.4
$\Delta\sigma(cp)$	154.1	148.5	107	80.6	222.9	195
$\Delta\sigma(ss)$	11	11	11.7	11	11	11.7
$\sigma = \sum_{i=1}^{n} \sigma_i$	591.7	651	1114	556.2	741.5	2178.1

The dependences of the total yield stress of steel of 100-meter differentially hardened rails on the distance to the railhead shown in Fig. 8.12 make it possible to trace the change in the strength of the metal at the yield point during the operation. It is seen that only the surface layer of the rail metal with a thickness of no more than 2 mm is subjected to strengthening. At a greater distance from the railhead surface, the strength properties of steel remain at the level of strength properties of steel in the initial state. The strength of the surface layer of steel essentially depends on the location of the analyzed layer, namely, the strength of the surface layer of the metal of the working fillet (Fig. 8.12, curve 1) is almost two times higher than the strength of the surface layer of the metal of the rolling surface (Fig. 8.12, curve 2).

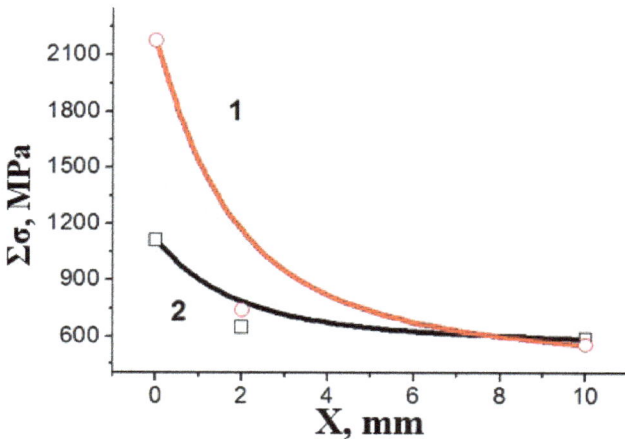

Figure 8.12 *Dependences of the total yield strength of steel of 100-meter differentially hardened rails on the distance to the railhead surface along the axis of symmetry of the working fillet (curve 1) and along the central axis (curve 2).*

8.5 Comparative analysis of the structural-phase state and hardening mechanisms at the yield point after the hauled load of 691.8 and 1411 million gross tons

In previous Chapters (3, 4, and 6) of this monograph, the results obtained in the analysis of the structure and phase composition of the metal of heat-strengthened rails both in the initial state and after operating load of 691.8 million gross tons during field tests are considered in detail. Studies carried out by the methods of transmission electron diffraction microscopy of thin foils have shown that plastic deformation of the metal during the operation is accompanied mainly by the transformation of the structure of pearlite lamellar morphology. The simultaneous occurrence of two processes of transformation of the structure and phase composition of lamellar pearlite colonies was discovered: firstly, the cutting of cementite plates (Fig. 8.13, a) and, secondly, the dissolution of cementite plates (Fig. 8.13, b).

The first process, which is carried out by the mechanism of cutting carbide particles and pulling away their fragments, is accompanied only by a change in their linear dimensions and morphology. After the hauled load of 691.8 million gross tons, this mechanism of destruction of pearlite colonies is traced mainly in the surface layer of the railhead (Fig. 8.14, d, e). After the hauled load of 1411 million gross tons, the destruction of cementite plates of pearlite colonies was found both in the surface layer of the railhead and at a distance of about 10 mm from it.

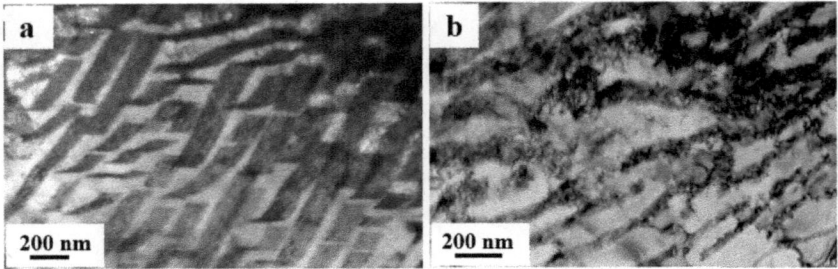

Figure 8.13 Electron microscopic images of a colony of lamellar pearlite after the hauled load of 1411 million gross tons (a) and 691.8 million gross tons (b).

The second process of destruction of cementite plates of pearlite colonies is carried out by the escape of carbon atoms from the crystal lattice of cementite to dislocations with the subsequent precipitation of nanosized particles of the carbide phase at the sub-boundaries and elements of the dislocation substructure. Photomicrographs shown in Figs. 8. 13(b) and 8.14(a) illustrate one of the stages of formation of a dislocation substructure around the cementite plates.

The studies performed showed that after the hauled load of 1411 million gross tons, the formation of a submicro-nanocrystalline subgrain (fragmented) structure is observed in the surface layer of the railhead (Fig. 8.15). In the layer that forms the surface of the working fillet, the sizes of subgrains (fragments) vary within 30 - 40 nm (Fig. 8.15, b). In the layer that forms the rolling surface, the sizes of subgrains (fragments) vary within 150 - 300 nm (Fig. 8.15, c). The relative content of subgrain (fragmented) structure in the surface layer of the working fillet is 0.25; in the top layer of the rolling surface is 0.15. After the hauled load of 691.8 million gross tons, the areas of material with a similar submicro-nanocrystalline subgrain structure are very rarely found in the metal structure.

Structure and Properties of Lengthy Rails after Extreme Long-Term Operation
Materials Research Foundations **106** (2021)

Materials Research Forum LLC
https://doi.org/10.21741/9781644901472

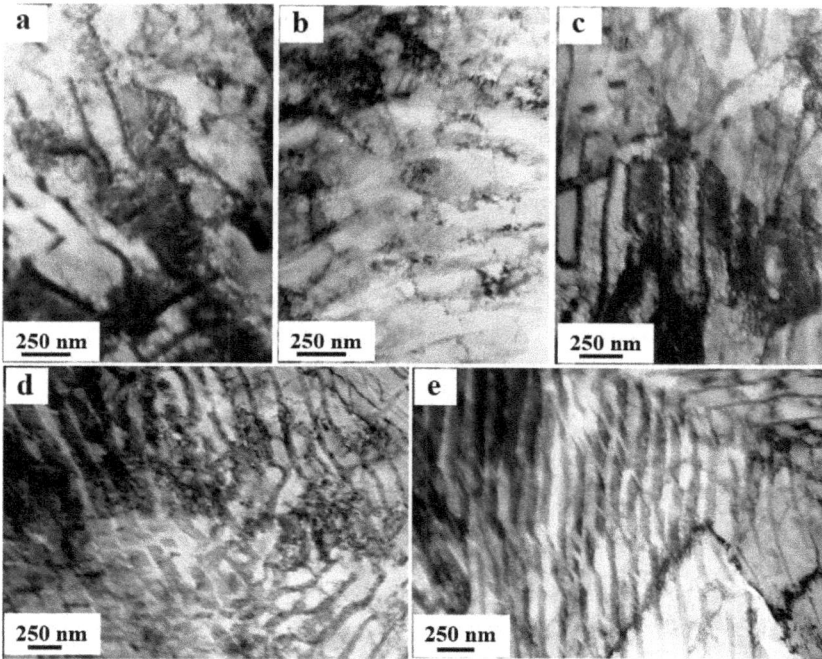

Figure 8.14 *Electron microscopic images of the structure forming after 691.8 million gross tons in the surface layer and the rolling surface (a-c) of the working fillet (d, e).*

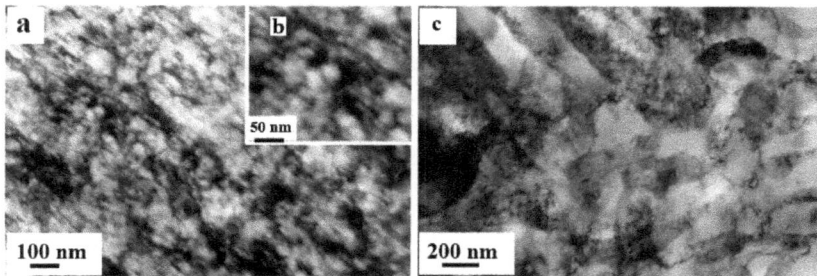

Figure 8.15 *Electron microscopic images of the structure formed after the hauled load of 1411 million gross tons in the surface layer of the working fillet (a, b) and the rolling surface (c).*

Thus, the results obtained indicate a significant increase in the transformation of the structure of the differentiated heat-strengthened metal of the railhead with an increase in the hauled tonnage. The discovered transformations of the steel structure will have a significant effect on the strength and plastic characteristics of the metal, ultimately determining the service life of the product. Evaluation of hardening mechanisms can help us to determine the regularities connecting the parameters of the structure and the strength properties of the material as well as to reveal the physical nature of the process of evolution of the properties.

Using the results of a quantitative analysis of the structure of steels obtained earlier in [141-144, 150, 155, 156, 212, 227, 231, 233-235, 263, 264], as well as in this monograph, the mechanisms of hardening were evaluated differentially hardened steel in the initial condition and after the hauled load of 691.8 million gross tons. The results of assessing the mechanisms of strengthening the structure of the metal rail, which is formed at various distances along the central axis and the axis of symmetry of the fillet in the head of 100-meter differentially hardened rails of the initial state are given in Table. 8.6, and after the hauled load of 691.8 million gross tons, Table. 8.7.

Table 8.6 *Estimates of the mechanisms of strengthening of the structure formed at different distances along the central axis and the axis of symmetry of the fillet in the head of 100-meter differentially hardened rails (initial state). The contributions are given taking into account the volume fraction of the structure with this hardening mechanism.*

Parameters, average by material, MPa	Rolling surface			Working fillet		
	Distance from the surface, mm					
	10	2	0	10	2	0
$\Delta\sigma(P)$	146	145	95.8	157	129	103.6
$\Delta\sigma(L)$	0	0	56	0	0	23
$\Delta\sigma(\rho)$	203	206	223	201	208	223
$\Delta\sigma(h)$	168	169	199.5	185.7	171.9	197.8
$\Delta\sigma(cp)$	103	114.4	167.2	119.4	177.3	182
$\Delta\sigma(ss)$	11	11	11.8	11	11	11.8
$\sigma = \sum_{i=1}^{n} \sigma_i$	631	645.4	753	684	697	741.4

The dependences of the total yield strength of steel of 100-meter differentially hardened rails on the distance to the railhead shown in Fig. 8.16 allow tracing the change in the strength of the metal at the yield point during the operation. It is seen that an increase in the hauled gross tonnage in the range of 691.8 - 1411 million tons leads to a significant (1.5-2 times) increase in the steel strength at the yield point. In this case, only the surface layer of the rail metal with a thickness of no more than 2 mm is subjected to strengthening. At a greater distance from the railhead surface, the strength properties of steel remain at the level of strength properties of steel in the initial state. The strength of the surface layer of steel depends significantly on the location of the analyzed layer, namely, the strength of the surface layer of the metal of the working fillet (Fig. 8.16, b) is almost 2 times higher than the strength of the surface layer of the metal of the rolling surface (Fig. 8.16, a).

Table 8.7 *Estimates of the mechanisms of strengthening of the structure formed at different distances along the central axis and the axis of symmetry of the fillet in the head of 100-meter differentially hardened rails after the long-term operation (hauled load of 691.8 million gross tons). The contributions are given taking into account the volume fraction of the structure with this hardening mechanism.*

Parameters, average by material, MPa	Rolling surface			Working fillet		
	Distance from the surface, mm					
	10	2	0	10	2	0
$\Delta\sigma(P)$	165	140	41	165	115	48
$\Delta\sigma(L)$	0	0	0	0	0	0
$\Delta\sigma(\rho)$	340	356	363	330	350	375
$\Delta\sigma(h)$	274	351	356	230	300	320
$\Delta\sigma(cp)$	0	0	113	0	0	67
$\Delta\sigma(ss)$	0	0	133	0	0	133
$\sigma = \sum_{i=1}^{n} \sigma_i$	779	847	1006	725	765	943

It was shown in papers [259, 260] and several other publications that plastic deformation of most metals and alloys is accompanied by fragmentation of the grain structure. The evolution of a fragmented structure proceeding with an increase in the degree of deformation leads to the formation of local areas that are incapable of further evolution (the so-called critical structure is formed [259-262]). Such a critical structure is the site of initiation of ductile fracture of the material. Based on the results shown in Tables 8.3 and 8.5, it can be expected that the destruction of the rail metal will primarily occur in the surface layer of the working fillet, where, after 1411 million tons, the formation of a subgrain structure with subgrain sizes within 100 nm is observed.

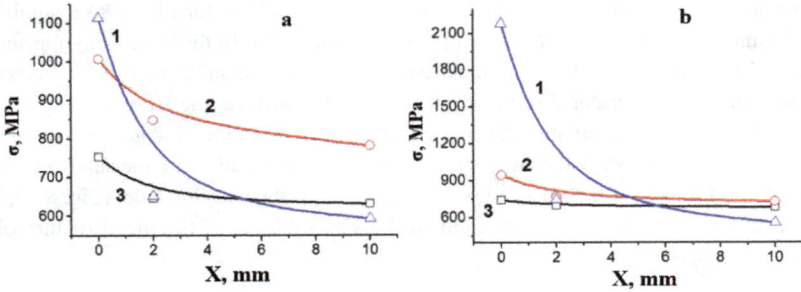

Figure 8.16 *Dependences of the total yield strength of steel of 100-meter differentiated hardened rails from the distance to the head surface along the central axis (a) and the axis of symmetry of the working fillet (b); (curves) 1: 1411 million tons; 2: 691.8 million tons; 3: initial state.*

Conclusion

The performed mechanical and tribological tests, the results of which are presented in this chapter, lead to the following conclusions:

1. In terms of impact strength and hardness on the rolling surface of the head and its cross-section, the metal of the investigated fragment meets the requirements of TU 0921-276-01124323-2012 for rails of the R350HT category. The hardness measured by the Rockwell method at a depth of 2 mm from the surface is 38.5 - 37.1 HRC, and at a depth of 10 and 22 mm is 34.8 - 35.8 HRC.

2. Microhardness at a depth of 2 mm from the rolling surface is 1481 - 1486 MPa. At a depth of up to 10 mm, the microhardness values decrease to 1210 - 1385 MPa, which is due to an increase in the inter-plate distance and a decrease in the level of strain hardening of the metal that occurs during the rail operation.

3. The operation of rails is accompanied by significant (1.5-2 times) strengthening of the near-surface layer of steel with a thickness of 80 - 100 microns, regardless of the studied section of the rail (rolling surface or working fillet). An increase in the wear resistance of the rolling surface by approximately 1.7 times to the volume of the product is demonstrated.

Plastic deformation of the metal, which occurs during the operation of the rail, is accompanied by the transformation of the structure of the lamellar pearlite colonies, firstly, by cutting and, secondly, dissolving the cementite plates. It is shown that these mechanisms of destruction of pearlite colonies can be traced in the surface layer of the rail head up to 10 mm thick. It was found that after the hauled load of 1411 million gross tons, a submicro-nanocrystalline subgrain (fragmented) structure is formed in the surface layer of the railhead. In the layer that forms the surface of the working fillet, the sizes of subgrains (fragments) vary within 30 - 40 nm. In the layer that forms the rolling surface, the sizes of subgrains (fragments) vary from 150 to 300 nm. The relative content of subgrain (fragmented) structure in the surface layer of the working fillet is 0.25; in the surface layer of the rolling surface is 0.15.

The estimates of the mechanisms of steel hardening at the yield point were carried out according to widely used in special literature well-tested ratios. It was shown that, firstly, the strength of steel is a multifactorial value and is determined by the combined action of several physical mechanisms. Secondly, the strength of the metal increases as it approaches the railhead surface. Third, the main mechanism of strengthening the rail metal in the surface layer (in the layer that forms the surface of the railhead) is the substructural one, caused by the interaction of moving dislocations with low-angle boundaries of fragments and subgrains. Fourth, an increase in the gross tonnage hauled in the range of 691.8 to 1411 million tons leads to a significant (1.5-2 times) increase in steel strength at the yield point. Fifth, the strength of the surface layer of the metal of the working fillet is almost two times higher than the strength of the surface layer of the metal of the rolling surface. It has been suggested that the destruction of the rail metal will primarily occur in the surface layer of the working fillet, where, after 1411 million tons, the formation of local sections with the so-called critical structure incapable of further evolution is observed.

Structure and Properties of Lengthy Rails after Extreme Long-Term Operation
Materials Research Foundations **106** (2021)

Materials Research Forum LLC
https://doi.org/10.21741/9781644901472

About the authors

Yuriev Anton Alekseevich (Yuriev A. A.) – Candidate of Technical Sciences, Product and Resource Management Manager of JSC EVRAZ ZSMK. *Research interests:* solid state physics, physical materials science. The results of scientific research are presented in more than 50 publications.

Gromov Viktor Evgenievich (Gromov V. E.) – Doctor of Physical and Mathematical Sciences, Professor, Head of the Department of Natural Sciences, Professor V.M. Finkel Siberian State Industrial University. *Research interests:* physical materials science, physics of strength and plasticity of materials under conditions of external energy influences. The results of scientific research are presented in more than 3700 publications.

Ivanov Yurii Fedorovich (Ivanov Yu. F.) – Doctor of Physical and Mathematical Sciences, Professor, and Chief Researcher at the Institute of High-Current Electronics SB RAS. *Research interests:* physical materials science, physics of strength and plasticity of materials under conditions of external energy influences, modification of inorganic materials by charged particle beams and plasma flows. The results of scientific research are presented in more than 3000 publications.

Rubannikova Yuliya Andreevna (Rubannikova Yu. A.) – Postgraduate student of the Department of Natural Sciences, Professor V.M. Finkel Siberian State Industrial University. *Research interests:* strength and plasticity of materials under conditions of external energy influences. The results of scientific research are presented in more than 60 publications.

Starostenkov Mikhail Dmitrievich (Starostenkov M. D.) – Doctor of Physical and Mathematical Sciences, Professor, Head of the Department of Physics at the I.I. Polzunov Altai State Technical University. Research interests: fundamental questions of materials science and condensed matter physics, evolution of defective structures in metals and alloys by computer simulation.

Tabakov Pavel Yaroslavovich (Tabakov P. Y.) – Professor, Department of Mechanical Engineering, Durban University of Technology, South Africa. *Research interests:* solid mechanics, composite materials, artificial intelligence, optimization. The results of scientific research are published in more than 80 international journals, conference proceedings and book chapters.

References

[1] Gromov V.E., Peregudov O.A., Ivanov Yu.F., Konovalov S.V., Yuriev A.A. Evolution of structural-phase states of rail metal during long-term operation. Novosibirsk: Publishing house of the Siberian Branch of the Russian Academy of Sciences, 2017, 164p.

[2] Gromov V.E., Morozov K.V., Peregudov O.A., Ivanov Yu.F. Formation of the microstructure of rails during hardening and long-term operation. Novokuznetsk: Publishing house of Siberian State Industrial University, 2017, 373p.

[3] Gromov V.E., Ivanov Yu.F., Yuriev A.B., Morozov K.V. Microstructure of quenched rails. Cambridge. CISP Ltd, 2016, 153p.

[4] Ivanisenko Yu., MacLaren I., Sauvage X., Valiev R.Z., Fecht H.-J. Shear-induced $\alpha{\rightarrow}\gamma$ transformation in nanoscale Fe-C composite. Acta Materialia, 2006, 54, pp. 1659-1669. https://doi.org/10.1016/j.actamat.2005.11.034

[5] Ning Jiang-li, Courtois-Manara E., Kurmanaeva L., Ganeev A.V., Valiev R.Z., Kübel C., Ivanisenko Yu. Tensile properties and work hardening behaviors of ultrafine grained carbon steel and pure iron processed by warm high pressure torsion. Materials Science and Engineering: A, 2013, 581, pp. 8-15. https://doi.org/10.1016/j.msea.2013.05.008

[6] Ivanisenko Yu., Wunderlich R.K., Valiev R.Z., Fecht H.-J. Annealing behavior of nanostructured carbon steel produced by severe plastic deformation. Scripta Materialia, 2003, 49(10), pp. 947-952. https://doi.org/10.1016/S1359-6462(03)00478-0

[7] MacLaren I., Ivanisenko Yu., Fecht H.-J., Sauvage X., Valiev R.Z. Early stages of nanostructuring of a pearlitic steel by high pressure torsion deformation. Ultrafine Grained Materials IV. Edited by Zhu E.T., et al. The Minerals, Metals & Materials Society, 2006, pp. 1-6.

[8] Ivanisenko Yu., Lojkowski W., Fecht H.-J. Stress- and Strain Induced Phase Transformations in Pearlitic Steels. Materials Science Forum, 2007, 539-543, pp. 4681-4686. https://doi.org/10.4028/www.scientific.net/MSF.539-543.4681

[9] Ivanisenko Yu., W. Lojkowski, Valiev R.Z., Fecht H.-J. The mechanism of formation of nanostructure and dissolution of cementite in a pearlitic steel during high pressure torsion. Acta Materialia, 2003, 51(18), pp. 5555-5570. https://doi.org/10.1016/S1359-6454(03)00419-1

[10] Meyers M.A., Mishra A., Benson D.J. Mechanical properties of nanocrystalline materials. Progress in Materials Science, 2006, 51, pp. 427-556. https://doi.org/10.1016/j.pmatsci.2005.08.003

[11] Kumar K.S., Swygenhoven H. Van, Suresh S. Mechanical behavior of nanocrystalline metals and alloys. Acta Materialia, 2003, 51, pp. 5743-5774. https://doi.org/10.1016/j.actamat.2003.08.032

[12] Lojkowski W., Millman Y., Chugunova S.I., Goncharova I.V., Djahanbakhsh M., Bürkle G., Fecht H.-J. The mechanical properties of the nanocrystalline layer on the surface of railway tracks. Materials Science and Engineering: A, 2003, 303(1-2), pp. 209-215. https://doi.org/10.1016/S0921-5093(00)01948-1

[13] Andrievsky R.A., Glezer A.M. Strength of nanostructures. Advances in Physical Sciences, 2009, 179(4), pp. 337-358. https://doi.org/10.3367/UFNr.0179.200904a.0337

[14] Koneva N.A., Kozlov E.V. Physical nature of the stages of plastic deformation. Structural Levels of Plastic Deformation and Fracture. Edited by Panina V.E. Novosibirsk: Nauka, 1990, pp. 123-186.

[15] Gapontsev V.L., Kondratyev V.V. Diffusion phase transformations in nanocrystalline alloys under severe plastic deformation. Reports of the Russian Academy of Sciences, 2002, 385(5), p. 684.

[16] Andrievsky R.L. Nanostructures in extreme conditions. Advances in Physical Sciences, 2014, 184(10), pp. 1017-1032. https://doi.org/10.3367/UFNr.0184.201410a.1017

[17] Lojkowski W., Djahanbakhsh M., Bürkle G., Gierlotka S., Zielinski W., Fecht H.-J. Nanostructure formation on the surface of railway tracks. Materials Science and Engineering: A, 2001, 303, pp.197-208. https://doi.org/10.1016/S0921-5093(00)01947-X

[18] Ivanisenko Yu., Fecht H.J. Microstructure modification in the surface layers of railway rails and wheels: effect of high strain rate deformation. Steel Tech, 2008, 3(1), pp. 19-23.

[19] Wu J., Petrov R., Kölling, S., Koenraad P., Malet L., Godet S., Sietsma J. Micro and Nanoscale Characterization of Complex Multilayer-Structured White Etching Layer in Rails. Metals, 2018, 8(10), pp. 749-761. https://doi.org/10.3390/met8100749

[20] Baumann G., Fecht H.J., Liebelt S. Formation of white etching layers on rail treads. Wear, 1996, 191, pp. 133-140. https://doi.org/10.1016/0043-1648(95)06733-7

[21] Osterle Rooch H., Pyzalla A., Wang LW., OeSterle W., Roach H., Pyzalla A. Wang L., Österle W., Roach H., Pyzalla A., et al. Investigation of white etching layers of rails by optical microscopy, electronmicroscopy, X-ray and synchrotron X-ray diffraction. Materials Science and Engineering A, 2001, 303, pp. 150-157. https://doi.org/10.1016/S0921-5093(00)01842-6

[22] Wild Wang, L., Hasse B., Wroblewski T., Goerigk G., Pyzalla A.E. Microstructure alterations at the surface of a heavily corrugated rail with strong ripple formation. Wear, 2000, 254, pp. 876-883. https://doi.org/10.1016/S0043-1648(03)00239-4

[23] Zhang H.W., Ohsaki S., Mitao S., Ohnuma M., Hono K. Microstructural investigation of white etching layer on pearlite steel rail. Materials Science and Engineering A, 2006, 421, pp. 191-199. https://doi.org/10.1016/j.msea.2006.01.033

Structure and Properties of Lengthy Rails after Extreme Long-Term Operation
Materials Research Foundations **106** (2021)

Materials Research Forum LLC
https://doi.org/10.21741/9781644901472

[24] Takahashi J., Kawakami K., Ueda M. Atom probe tomography analysis of the white etching layer in a railtrack surface. Acta Materialia, 2010, 58, pp. 3602-3612. https://doi.org/10.1016/j.actamat.2010.02.030

[25] Lojkowski W., Djahanbakhsh M., Bürkle, G., Gierlotka S., Zielleski W., Pecht H. J. Nanostructure formation on the surface of railway tracks. Materials Science and Engineering A, 2001, 303, pp. 197-208. https://doi.org/10.1016/S0921-5093(00)01947-X

[26] Newcomb S.B., Stobbs W.M. A transmission electron microscopy study of the white-etching layer on a railhead. Materials Science and Engineering, 1984, 66, pp. 195-204. https://doi.org/10.1016/0025-5416(84)90180-0

[27] Ishida, M. Rolling contact fatigue (RCF) defects of rails in Japanese railways and its mitigation strategies. Electronic Journal of Structural Engineering, 2013, 13, pp. 67-74.

[28] Steenbergen M., Dollevoet R. On the mechanism of squat formation on train rails: Part I. Origination. International Journal of Fatigue, 2013, 47, pp. 361-372. https://doi.org/10.1016/j.ijfatigue.2012.04.023

[29] Pal S., Valente C., Daniel W., Farjoo M. Metallurgical and physical understanding of rail squat initiation and propagation. Wear, 2012, 284, pp. 30-42. https://doi.org/10.1016/j.wear.2012.02.013

[30] Clayton P. Tribological aspects of wheel-rail contact: A review of recent experimental research. Wear, 1995, 191, pp. 170-183. https://doi.org/10.1016/0043-1648(95)06651-9

[31] Carroll R.J, Beynon J.H. Rolling contact fatigue of white etching layer: Part 1. Crack morphology. Wear, 2007, 262, pp. 1253-1266. https://doi.org/10.1016/j.wear.2007.01.003

[32] Carroll R.I., Beynon J.H. Rolling contact fatigue of white etching layer: Part 2. Numerical results. Wear, 2007, 262, pp. 1267-1273. https://doi.org/10.1016/j.wear.2007.01.002

[33] Wang L., Pyzalla A., Stadlbauer W., Werner E. A. Microstructure features on rolling surfaces of railway rails subjected to heavy loading. Materials Science and Engineering A, 2003, 359, pp. 31-43. https://doi.org/10.1016/S0921-5093(03)00327-7

[34] Lojkowski W., Millman Y., Chugunova, S.I., Goncharova I.V., Djahanbakhsh M., Bürkle G., Fecht, H.J. The mechanical properties of the nanocrystalline layer on the surface of railway tracks. Materials Science and Engineering A, 2001, 303, pp. 209-215. https://doi.org/10.1016/S0921-5093(00)01948-1

[35] Wu J., Petrov R.H., Naeimi M., Li, Z., Dollevoet R., Sietsma J. Laboratory simulation of martensite formation of white etching layer in rail steel. International Journal of Fatigue, 2016, 91, pp. 11-20. https://doi.org/10.1016/j.ijfatigue.2016.05.016

Structure and Properties of Lengthy Rails after Extreme Long-Term Operation
Materials Research Foundations **106** (2021)

Materials Research Forum LLC
https://doi.org/10.21741/9781644901472

[36] Griffiths B.J. White layer formations at machined surfaces and their relationship to white layer Formations at worn surfaces. Journal of Tribology, 1985, 107, pp. 165-170. https://doi.org/10.1115/1.3261015

[37] Umbrello D., Rotella G. Experimental analysis of mechanisms related to white layer formation during hard turning of AISI 52100 bearing steel. Materials Science and Technology, 2012, 28, pp. 205-212. https://doi.org/10.1179/1743284711Y.0000000020

[38] Todaka Y., Umemoto M., Tsuchiya K. Nanocrystallization in carbon steels by various severe plastic deformation processes. Nanomaterials by severe plastic deformation; Wiley-VCH Verlag GmbH & Co. KGaA: Weinheim, 2004, pp. 505-510. https://doi.org/10.1002/3527602461.ch9d

[39] Rauch E.F., Véron M. Automated crystal orientation and phase mapping in TEM. Materials Characterization, 2014, 98, pp. 1-9. https://doi.org/10.1016/j.matchar.2014.08.010

[40] Kobler A., Kashiwar A., Hahn H., Kübel C. Combination of in situ straining and ACOM TEM: A novel method for analysis of plastic deformation of nanocrystalline metals. Ultramicroscopy, 2013, 128, pp. 68-81. https://doi.org/10.1016/j.ultramic.2012.12.019

[41] Linz M., Cihak-Bayr U., Trausmuth A., Scheriau S., Künstner D., Badisch E. EBSD study of early-damaging phenomena in wheel-rail model test. Wear, 2015, 342, pp. 13-21. https://doi.org/10.1016/j.wear.2015.08.004

[42] Wu J., Petrov R.H., Naeimi M., Li Z., Sietsma J.A Microstructural study of rolling contact fatigue in rails. Proceedings of the Second International Conference on Railway Technology: Research, Development and Maintenance, Ajaccio, 2014, p. 118.

[43] Hossain R., Pahlevani F., Witteveen E., Banerjee A., Joe B., Prusty B.G., Dippenaar R., Sahajwalla V. Hybrid structure of white layer in high carbon steel-formation mechanism and its properties. Scientific Reports, 2017, 7, pp. 1-12. https://doi.org/10.1038/s41598-017-13749-7

[44] Bernsteiner C., Muller G., Meierhofer A., et al. Development of white etching layers on rails: Simulation and experiments. Wear, 2016, 366-367, pp. 116-122. https://doi.org/10.1016/j.wear.2016.03.028

[45] Peregudov O.A. Changes in the fine structure and properties of rails during long-term operation. Abstract of the dissertation for the degree of candidate of technical sciences. Novokuznetsk: Publishing house of Siberian State Industrial University, 2017, 12p.

[46] Ivanov Yu.F., Gromov V.E., Peregudov O.A., Morozov K.V., Yuriev A.B. Evolution of the structural-phase states of rails during long-term operation. Bulletin of Higher Educational Institutions. Ferrous Metallurgy, 2015, 58(4), pp. 262-267.

[47] Peregudov O.A., Morozov K.V., Gromov V.E., Glezer A.M., Ivanov Yu.F. Formation of fields of internal stresses in rails during long-term operation. Deformation and Destruction of Materials, 2015, No. 11, pp. 34-37.

[48] Gromov V.E., Ivanov Yu.F., Morozov K.V., Peregudov O.A., Alsaraeva K.V., Popova N.A., Nikonenko E.L. Changes in the structure and properties of the surface layers of the rail head after long-term operation. Fundamental'nye problemy sovremennogo materialovedenia (Basic Problems of Material Science), 2015, 12(2), pp. 203-208.

[49] Ivanov Yu.F., Morozov K.V., Peregudov O.A., Gromov V.E., Popova N.A., Nikonenko E.L. Formation of structural-phase gradients in rails during long-term operation. Problems of Ferrous Metallurgy and Material Science, 2015, No. 3, pp. 59-64.

[50] Gromov V.E., Ivanov Yu.F., Morozov K.V., Peregudov O.A., Popova N.A., Nikonenko E.L. Mechanisms of rail hardening during long-term operation. Problems of Ferrous Metallurgy and Material Science, 2015, No. 4, pp. 98-104.

[51] Gromov V.E., Peregudov O.A., Ivanov Yu.F., Morozov K.V., Alsaraeva K.V. Evolution of the structure and properties of the surface layer of rails during long-term operation. Problems of Materials Science, 2015, 3(83), pp. 30-38.

[52] Morozov K.V., Gromov V.E., Peregudov O.A., Ivanov Yu.F., Yuriev A.B., Aksenova K.V. Formation of the fine structure of rails during volumetric and differentiated hardening. Problems of Ferrous Metallurgy and Material Science, 2016, No. 1, pp. 53-61.

[53] Ivanov Yu.F., Morozov K.V., Peregudov O.A., Gromov V.E. Operation of rail steel: degradation of the structure and properties of the surface layer. Bulletin of Higher Educational Institutions. Ferrous Metallurgy, 2016, 59(8), pp. 576-580. https://doi.org/10.17073/0368-0797-2016-8-576-580

[54] Ivanov Yu.F., Gromov V.E., Glezer A.M., Peregudov O.A., Morozov K.V. The nature of degradation of the structure of the rolling surface of rails during operation. Proceedings of Russian Academy of Sciences. Physical series, 2016, 80(12), pp. 1682-1687.

[55] Gromov V.E., Peregudov O.A., Ivanov Yu.F., Morozov K.V., Alsaraeva K.V., Semina O.A. Surface layer structure degradation of rails in prolonged operation. Journal of Surface Investigation. X-ray, Synchrotron and Neutron Techniques, 2015, 9(6), pp. 1292-1298.

[56] Gromov V.E., Ivanov Yu.F., Peregudov O.A., Morozov K.V., Wang X.L., Dai W.B., Ponomareva Yu.V., Semina O.A. Evolution of structure and properties of railhead fillet in long-term operation. Materials and Electronics Engineering, 2015, 2(4), pp. 1-4. https://doi.org/10.11605/mee-2-4

[57] Peregudov O.A., Gromov V.E., Ivanov Yu.F., Morozov K.V., Alsaraeva K.V., Semina O.A. Structure-phase states evolution in rails during long operation. AIP Conference Proceedings, 2015, 1683(020179), 4 p. https://doi.org/10.1063/1.4932869

[58] Ivanov Yu.F., Peregudov O.A., Morozov K.V., Gromov V.E., Popova N.A., Nikonenko E.N. Formation structural phase gradients in rail steel during long-term operation. IOP Conference

Structure and Properties of Lengthy Rails after Extreme Long-Term Operation
Materials Research Foundations **106** (2021)

Materials Research Forum LLC
https://doi.org/10.21741/9781644901472

Series: Materials Science and Engineering, 2016, 112(012038), 4 p. https://doi.org/10.1088/1757-899X/112/1/012038

[59] Gromov V.E., Morozov K.V., Ivanov Yu.F., Aksenova K.V., Peregudov O.A., Semin A.P. Formation and evolution of structure-phase states in rails after drawn resource. Diagnostics, Resource and Mechanics of Materials and Structures, 2016, No. 1, pp. 38-44. https://doi.org/10.17804/2410-9908.2016.1.038-044

[60] Gromov V.E., Peregudov O.A., Ivanov Y.F., Glezer A.M., Morozov K.V., Aksenova K.V., Semina O.A. Physical nature of rail strengthening in long term operation. AIP Conference Proceedings, 2016, 1783(020069), 4 p. https://doi.org/10.1063/1.4966362

[61] Dylewski B., Risbet M., Bouvier S. The three-dimensional gradient of microstructure in worn rails. Experimental characterization of plastic deformation accumulated by RCF. Wear, 2017, 392-393, pp. 50-59. https://doi.org/10.1016/j.wear.2017.09.001

[62] Neslušan M., Čížek J., Zgútová K., Kejzlar P., Šramek J., Čapek J., Hruška P., Melikhova O. Microstructural transformation of a rail surface induced by severe thermoplastic deformation and its non-destructive monitoring via Barkhausen noise. Wear, 2018, 402-403, pp. 38-48. https://doi.org/10.1016/j.wear.2018.01.014

[63] Lewis R., Christoforou P., Wang W.J., Beagles A., Burstow M., Lewis S.R. Investigation of the influence of rail hardness on the wear of rail and wheel materials under dry conditions (ICRI wear mapping project). Wear, 2019, 430-431, pp. 383-392. https://doi.org/10.1016/j.wear.2019.05.030

[64] Kalousek J., Fegredo D.M., Laufer E.E. The wear resistance and worn metallography of pearlite, bainite and tempered martensite rail steel microstructures of high hardness. Wear, 1985, 105, pp. 199-222. https://doi.org/10.1016/0043-1648(85)90068-7

[65] Wang M., Zhang C., Sun D., Yang Z., Zhang F. Wear behaviour and microstructure evolution of pearlitic steels under block-on-ring wear process. Materials Science and Technology, 2019, 35(8), pp. 1149-1160. https://doi.org/10.1080/02670836.2019.1613043

[66] Roy T., Lai Q., Abrahams R., Mutton P., Paradowska A., Soodi M., Yan W. Effect of deposition material and heat treatment on wear and rolling contact fatigue of laser cladded rails. Wear, 2018, 412-413, pp. 69-81. https://doi.org/10.1016/j.wear.2018.07.001

[67] Thakkar N.A., Steel J.A., Reuben R.L. Rail-wheel interaction monitoring using Acoustic Emission: A laboratory study of normal rolling signals with natural rail defects. Mechanical Systems and Signal Processing, 2010, 24(1), pp. 256-266. https://doi.org/10.1016/j.ymssp.2009.06.007

[68] Wen Z., Jin X., Xiao X., et al. Effect of a scratch on curved rail on initiation and evolution of plastic deformation induced rail corrugation. International Journal of Solids and Structures, 2008, 45(7-8), pp. 2077-2096. https://doi.org/10.1016/j.ijsolstr.2007.11.013

[69] Jianxi W., Yude X., Songliang L., et al. Probabilistic prediction model for initiation of RCF cracks in heavy-haul railway. International Journal of Fatigue, 2011, 33(2), pp. 212-216. https://doi.org/10.1016/j.ijfatigue.2010.08.006

[70] Mohammadzadeh S., Sharavi M., Keshavarzian H. Reliability analysis of fatigue crack initiation of railhead in bolted rail joint. Engineering Failure Analysis, 2013, 29, pp. 132-148. https://doi.org/10.1016/j.engfailanal.2012.11.012

[71] Tarasov S. Y. and Rubtsov V. E. Shear instability in the subsurface layer of material in friction. Physics of Solid State, 2011, 53, pp. 358-362. https://doi.org/10.1134/S1063783411020302

[72] Tarasov S.Y., Rubtsov V.E. and Kolubaev A.V. Subsurface shear instability and nanostructuring of metals in sliding. Wear, 2010, 268, pp. 59-66. https://doi.org/10.1016/j.wear.2009.06.027

[73] Rubtsov V. E., Tarasov S. Y. and Kolubaev A. V. One-dimensional model of inhomogeneous shear in sliding. Physical Mesomechanics, 2012, 15, pp. 337-341. https://doi.org/10.1134/S1029959912030174

[74] Klassen-Nekludova M. V., Kontorova T. A. Nature of intergranular layers. Uspekhi Phizicheskikh Nauk, 1939, 22, pp. 249-292. https://doi.org/10.3367/UFNr.0022.193907a.0249

[75] Klassen-Nekludova M. V., Kontorova T. A. Nature of intergranular layers. Uspekhi Phizicheskikh Nauk, 1939, 22, pp. 395-426. https://doi.org/10.3367/UFNr.0022.193908c.0395

[76] Grabsky M.V. The structure of grain boundaries in metals. Monograph. Translated from Polish. Moscow: Metallurgy, 1972, 160p.

[77] Sundeev R. V., et al. Susceptibility of crystalline alloys to deformational amorphization during torsion under quasi-hydrostatic pressure. Bulletin of the Russian Academy of Sciences: Physics, 2012, 76, pp. 1226-1232. https://doi.org/10.3103/S1062873812110202

[78] Sundeev R. V., et al. Deformation behavior of layered amorphous-crystalline Ti-Ni-Cu composite under different conditions of torsion in a bridgman chamber. Bulletin of the Russian Academy of Sciences: Physics, 2015, 79, pp. 1156-1161. https://doi.org/10.3103/S1062873815090221

[79] Glezer A.M., Sundeev R. V. General view of severe plastic deformation in solid state. Materials Letters, 2015, 139, pp. 455-457. https://doi.org/10.1016/j.matlet.2014.10.052

[80] Sarychev V.D., Nevskii S.A., Sarycheva E.V., et al. Viscous flow analysis of the Kelvin-Helmholtz instability for short waves. AIP Conference Proceedings, 2016, 1783, pp. 020198 (1-4). https://doi.org/10.1063/1.4966492

[81] Sarychev V.D., Nevsky S.A., Gromov V.E. Model of the formation of nanostructures in rail steel under severe plastic deformation. Deformation and Destruction of Materials, 2016, No. 6, pp. 25-29.

[82] Sarychev V.D., Nevskii S.A., Granovskii A.Yu., et al. Viscous flow analysis of the Kelvin-Helmholtz instability for short waves. AIP Conference Proceedings, 2015, 1683, pp. 020200 (1-4). https://doi.org/10.1063/1.4966492

[83] Daves W., Kubin W., Scheriau S., et. al. A finite element model to simulate the physical mechanism of wear and crack initiation in wheel-rail contact. Wear, 2016, 366-367, pp. 78-83. https://doi.org/10.1016/j.wear.2016.05.027

[84] Kolubaev A., Tarasov S., Sizova O., Kolubaev E. Scale-dependent subsurface deformation of metallic materials in sliding. Tribology International, 2010, 43(4), pp. 695-699. https://doi.org/10.1016/j.triboint.2009.10.009

[85] Panin V.E., Kolubaev A.V., Tarasov S.Yu., Popov V.L. Subsurface layer formation during sliding friction. Wear, 2002, No. 10-11, pp. 860-867. https://doi.org/10.1016/S0043-1648(01)00819-5

[86] Rubtsov V.E., Tarasov S.Yu., Kolubaev A.V. One-dimensional model of inhomogeneous shear under sliding friction. Physical Mesomechanics, 2012, 15(4), pp. 103-108. https://doi.org/10.1134/S1029959912030174

[87] Tarasov S.Yu., Rubtsov V.E., Kolubaev A.V., Gorbatenko V.V. Analysis of macroscopic deformation fields during sliding friction. Bulletin of Higher Educational Institutions. Physics, 2013, 56(7-2), pp. 350-355.

[88] Rubtsov V.E., Tarasov S.Yu., Kolubaev A.V. Inhomogeneity of deformation and shear instability of material under friction. Bulletin of Higher Educational Institutions. Physics, 2011, 11(3), pp. 215-220.

[89] Monin A.S., Yaglom A.M. Statistical fluid mechanics. Vol. 1. Moscow: Nauka, 1967, 640 p.

[90] Landau L.D., Lifshits E.M. Theoretical physics. Vol. 6: Hydrodynamics. Moscow: Fizmatlit, 2006.

[91] Ghaffari M.A., Zhang Y., Xiao S. Multiscale modeling and simulation of rolling contact fatigue. International Journal of Fatigue, 2018, 108, pp. 9-17. https://doi.org/10.1016/j.ijfatigue.2017.11.005

[92] Wang J., Zhou C. Finite element analysis of the wear fatigue of rails with gradient structures. Materials Letters, 2018, 231, pp. 35-37. https://doi.org/10.1016/j.matlet.2018.08.012

Materials Research Forum LLC
https://doi.org/10.21741/9781644901472

[93] Kim D., Quagliato L., Park D., Kim N. Lifetime prediction of linear slide rails based on surface abrasion and rolling contact fatigue-induced damage. Wear, 2019, 420-421, pp 184-194. https://doi.org/10.1016/j.wear.2018.10.015

[94] Mazzù A., Donzella G. A model for predicting plastic strain and surface cracks at steady-state wear and ratcheting regime. Wear, 2018, 400-401, pp. 127-136. https://doi.org/10.1016/j.wear.2018.01.002

[95] Chen H., Ji Y., Zhang C., Liu W., Chen H., Yang Z., Chen L.-Q., Chen L. Understanding cementite dissolution in pearlitic steels subjected to rolling-sliding contact loading: A combined experimental and theoretical study. Acta Materialia, 2017, 141, pp. 193-205. https://doi.org/10.1016/j.actamat.2017.09.017

[96] Khvostik M.Yu., Khromov I.V., Bykova O.A., Beresten G.A. Analysis of the state of the working surface of rails of the pilot batches on the Experimental ring of JSC "VNIIZhT". Bulletin of the Scientific Research Institute of Railway Transport, 2018, 77(3), pp. 141-148. https://doi.org/10.21780/2223-9731-2018-77-3-141-148

[97] Kossov V.S., Volokhov G.M., Krasnov O.G., Ovechnikov M.N., Protopopov A.L., Oguenko V.V. Influence of the value of axial loads of rolling stock on the contact-fatigue life of rails. Bulletin of the Scientific Research Institute of Railway Transport, 2018, 77(3), pp. 149-156. https://doi.org/10.21780/2223-9731-2018-77-3-149-156

[98] Kogan A. Ya. Mathematical model of the emergence and development of wavy rail wear when an electric locomotive moves in traction mode in straight track sections. Bulletin of the Scientific Research Institute of Railway Transport, 2019, 78(3), pp. 131-140. https://doi.org/10.21780/2223-9731-2019-78-3-131-140

[99] Sakalo V.I., Sakalo A.V. Criteria for predicting the occurrence of contact fatigue damage in the wheels of railway rolling stock and rails. Bulletin of the Scientific Research Institute of Railway Transport, 2019, 78(3), pp. 141-148. https://doi.org/10.21780/2223-9731-2019-78-3-141-148

[100] Shur E.A., Fedin V.M., Borts A.I., Ronzhina Yu.V., Fimkin A.I. Ways to eliminate increased damage to rails in the zone of welded joints. Bulletin of the Scientific Research Institute of Railway Transport, 2019, 78(4), pp. 210-217. https://doi.org/10.21780/2223-9731-2019-78-4-210-217

[101] Markov D.P. On the dehision-deformation nature of friction and wear. Bulletin of the Scientific Research Institute of Railway Transport, 2019, 78(5), pp. 303-312. https://doi.org/10.21780/2223-9731-2019-78-5-312

[102] Zakharov S.M., Torskaya E.V. Approaches to modeling the occurrence of surface contact-fatigue damage in rails. Bulletin of the Scientific Research Institute of Railway Transport, 2018, 77(5), pp. 259-268. https://doi.org/10.21780/2223-9731-2018-77-5-259-268

Materials Research Forum LLC
https://doi.org/10.21741/9781644901472

[103] Gromov V.E., Yuriev A.B., Morozov K.V., Ivanov Yu.F. Microstructure of hardened rails. Novokuznetsk: Publishing house "Inter-Kuzbass", 2014, 213p.

[104] Chernyavsky V.S. Stereology in metal science. Moscow: Metallurgy, 1977, 280p.

[105] Saltykov S.A. Stereometric metallography. Moscow: Metallurgy, 1970, 376p.

[106] Jian Min Zuo, John C.H. Spence. Advanced Transmission Electron Microscopy. Springer, New York, 2017, 729 p.

[107] Fultz B., Howe J. Transmission Electron Microscopy and Diffractometry of Materials. Fourth edition. Berlin: Springer, 2013, 764p. https://doi.org/10.1007/978-3-642-29761-8

[108] Thomas J., Gemming T. Analytical Transmission Electron Microscopy. Dordrecht: Springer Netherlands, 2014, 348p. https://doi.org/10.1007/978-94-017-8601-0

[109] Egerton F.R. Physical Principles of Electron Microscopy. Basel: Springer International Publishing, 2016, 196p. https://doi.org/10.1007/978-3-319-39877-8_2

[110] Kumar C.S.S.R. Transmission Electron Microscopy. Characterization of Nanomaterials. New York: Springer, 2014, 717p. https://doi.org/10.1007/978-3-642-38934-4

[111] Carter C.B., Williams D.B. (Ed.). Transmission Electron Microscopy. Berlin: Springer International Publishing, 2016, 518p. https://doi.org/10.1007/978-3-319-26651-0

[112] Hirsch P., Howie A., Nicholson P., et al. Electron microscopy of thin crystals. Moscow: Mir, 1968, 574p.

[113] Utevsky L.M. Diffraction electron microscopy in metal science. Moscow: Metallurgy, 1973, 584p.

[114] Schumann H. Metallographie. Leipzig: VEB, 1964, 621p.

[115] Klopotov A.A., Abzaev Yu.A., Potekaev A.I., Volokitin O.G., Klopotov V.D. Physical foundations of X-ray structural study of crystalline materials. Tomsk: Publishing house of the Tomsk Polytechnic University, 2013, 263p.

[116] Koneva N.A., Lychagin D.V., Teplyakova L.A. Lattice rotations and stages of plastic deformation. Experimental Research and Theoretical Description of Disclinations, Leningrad: Institute of Physics and Technology, 1984, pp. 161-164.

[117] Koneva N.A., Lychagin D.V., Zhukovsky S.P., et al. Evolution of dislocation structure and stages of plastic flow of polycrystalline iron-nickel alloy. Physics of Metals and Metal Science, 1985, 60(1), pp. 171-179.

[118] Koneva N.A., Kozlov E.V. The nature of substructural hardening. Bulletin of Higher Educational Institutions. Physics, 1982, No. 8, pp. 3-14. https://doi.org/10.1007/BF00895238

Structure and Properties of Lengthy Rails after Extreme Long-Term Operation
Materials Research Foundations **106** (2021)

Materials Research Forum LLC
https://doi.org/10.21741/9781644901472

[119] Koneva N.A., Lychagin D.V., Teplyakova L.A., et al. Strip substructure in fcc-single-phase alloys. In the book: Disclinations and Rotational Deformation of Solids. Leningrad: Institute of Physics and Technology, 1988, pp. 103-113.

[120] Teplyakova L.A., Ignatenko L.N., Kasatkina N.F., et al. Regularities of plastic deformation of steel with a tempered martensite structure. In the book: Plastic Deformation of Alloys. Structurally Heterogeneous Materials. Tomsk: Tomsk State University, 1987, pp. 26-51.

[121] Ivanov Yu.F., Gromov V.E., Popova N.A., Konovalov S.V., Koneva N.A. Structural-phase states and mechanisms of hardening of deformed steel. Novokuznetsk: Polygraphist, 2016, 510p.

[122] Tribology. Physical foundations, mechanics and technical applications: Textbook for universities. Edited by Berkovich I.I. and Gromakovsky D.G. Samara: Samara State Technical University, 2000, 268p.

[123] Gromov V.E., Ivanov Yu.F., Yuriev A.A., Morozov K.V., Konovalov S.V. Differentiated hardened rails: evolution of structure and properties during operation. Novokuznetsk: Siberian State Industrial University, 2017, 197p.

[124] Gromov V.E., Volkov K.V., Ivanov Yu.F., Morozov K.V. et al. Formation of the fine structure of metal rails with increased wear resistance. Problems of Materials Science, 2013, 4(76), pp. 15-23.

[125] Volkov K.V., Gromov V.E., Ivanov Yu.F., Yuriev A.B., Morozov K.V., Alsaraeva K.V. Formation of structure, phase composition and fine substructure in differentially hardened rails. Fundamental'nye problemy sovremennogo materialovedenia (Basic Problems of Material Science), 2014, 11(1), pp. 50-55.

[126] Gromov V.E., Volkov K.V., Ivanov Yu.F., Morozov K.V., Konovalov S.V., Alsaraeva K.V. Structure, phase composition and defective substructure of rails of the highest quality category. Bulletin of Higher Educational Institutions. Physics, 2014, 2, pp. 72-76. https://doi.org/10.1007/s11182-014-0233-7

[127] Gromov V.E., Volkov K.V., Ivanov Yu.F., Yuriev A.B., Konovalov S.V., Morozov K.V. Formation of a fine structure in rails of low-temperature reliability. Problems of Ferrous Metallurgy and Material Science, 2013, 4, pp. 61-68.

[128] Ivanov Yu.F., Gromov V.E., Volkov K.V., Morozov K.V., et al. Formation of gradients of structure, phase composition and defective substructure in rails during differentiated hardening. Perspective Materials, 2014, 3, pp. 40-45.

[129] Gromov V.E., Ivanov Yu.F., Glezer A.M., Morozov K.V., et al. Differentiated rail hardening: structure, phase composition and defect substructure of the surface layer. Deformation and Destruction of Materials, 2014, 5, pp. 42-46.

Structure and Properties of Lengthy Rails after Extreme Long-Term Operation
Materials Research Foundations **106** (2021)

Materials Research Forum LLC
https://doi.org/10.21741/9781644901472

[130] Gromov V.E., Volkov K.V., Ivanov Yu.F., Morozov K.V., et al. Formation of dislocation substructure and internal stress fields in volumetric and differentiated hardened rails. Nanoengineering, 2014, 3(33), pp. 22-26.

[131] Gromov V.E., Morozov K.V., Ivanov Yu.F., Volkov K.V., et al. Formation of gradients of structure, phase composition and defect substructure in differentially hardened rails. Russian Nanotechnologies, 2014, 9(5-6), pp. 59-62. https://doi.org/10.1134/S1995078014030045

[132] Gromov V.E., Volkov K.V., Yuriev A.B., Morozov K.V., et al. Structural-phase states and defect substructure of differentially hardened rails. Bulletin of Higher Educational Institutions. Ferrous Metallurgy, 2014, 57(12), pp. 29-32. https://doi.org/10.17073/0368-0797-2014-12-29-32

[133] Gromov V.E., Volkov K.V., Ivanov Yu.F., Morozov K.V., et al. Formation of structure, phase composition and defective substructure in volumetric hardened rails of special categories. Bulletin of Higher Educational Institutions. Ferrous Metallurgy, 2014, 6, pp. 54-61. https://doi.org/10.17073/0368-0797-2014-6-54-61

[134] Ivanov Yu.F., Gromov V.E., Yuriev A.B., Volkov K.V., Morozov K.V., et al. Formation of internal stress fields in rails. Problems of Ferrous Metallurgy and Materials Science, 2014, 1, pp. 79-84.

[135] Gromov V.E., Volkov K.V., Glezer A.M., Ivanov Yu.F., Morozov K.V., et al. Dislocation substructure and internal stress fields in volumetric and differentiated hardened rails. Proceedings of Russian Academy of Sciences. Physical series, 2014, 78(10), pp. 1230-1237. https://doi.org/10.3103/S1062873814100086

[136] Gromov V.E., Morozov K.V., Ivanov Yu.F., Volkov K.V., et al. Structural-phase state of the surface layers of rails subjected to differentiated hardening. Metall Technology, 2014, 12, pp. 15-19.

[137] Gromov V.E., Volkov K.V., Ivanov Yu.F., Morozov K.V., Alsaraeva K.V., Konovalov S.V. Formation of structure, phase composition and defective substructure in volumetric and differentially hardened rails. Progress in Physics of Metals, 2014, 15(1), pp. 1-33.

[138] Gromov V.E., Yuriev A.B., Morozov K.V., Volkov K.V., Ivanov Yu.F. Comparative analysis of structural-phase states in rails after volumetric and differentiated hardening. Steel, 2014, No. 7, pp. 91-95.

[139] Morozov K.V., Gromov V.E., Ivanov Yu.F., Yuriev A.B., Bataev V.A. The nature of hardening of the grain structure of rails subjected to bulk hardening. Fundamental'nye problemy sovremennogo materialovedenia (Basic Problems of Material Science), 2014, 11(3), pp. 293-297.

[140] Morozov K.V., Gromov V.E., Ivanov Yu.F., Glazep A.M., Bataev V.A. Analysis of structural and phase states in rails subjected to volumetric and differential hardening. Factory Laboratory. Diagnostics of Materials, 2015, No. 4, pp. 22 - 26.

[141] Gromov V.E., Yuriev A.A., Morozov K.V., et al. Evolution of fine structure in surface layers of 100-m differentially hardened rails during long-term operation. Fundamental'nye problemy sovremennogo materialovedenia (Basic Problems of Material Science), 2017, 14(2), pp. 267-273.

[142] Yuriev A.A., Gromov V.E., Morozov K.V., et al. Changes in the structure and phase composition of the surface of 100-m differentially hardened rails during long-term operation. Bulletin of Higher Educational Institutions. Ferrous Metallurgy, 2017, 60(10), pp. 826-831. https://doi.org/10.17073/0368-0797-2017-10-826-830

[143] Ivanov Yu.F. Gromov V.E., Yuriev A.A. et al. The nature of surface hardening of differentially hardened rails during long-term operation. Deformation and Destruction of Materials, 2018, No. 4, pp. 67-85.

[144] Gromov V.E., Yuryev A.A., Ivanov Yu.F., et al. Evolution of the structure and properties of differentially hardened rails during long-term operation. Metal Physics and the Latest Technologies, 2017, 39(12), pp. 1599-1646.

[145] Gromov V.E., Gromov V.E., Yuriev A.A., Ivanov Yu.F., et al. Transformation of the structure of 100-meter differentially hardened rails during long-term operation. Fundamental'nye problemy sovremennogo materialovedenia (Basic Problems of Material Science), 2018, 15(1), pp. 128-134.

[146] Gromov V.E., Yuriev A.A., Peregudov O.A., etc. Physical nature of structure and properties degradation of rail surface after long term operation. Key Engineering Materials, AIP Conference Proceedings, 2017, 1909, p. 020066.

[147] Koneva N.A., Kozlov E.V. Physics of substructural hardening. Bulletin of Tomsk State University of Architecture and Civil Engineering, 1999, No. 1, pp. 21-35.

[148] Kozlov E.V., Starenchenko V.A., Koneva N.A. Evolution of dislocation substructure and thermodynamics of plastic deformation of metallic materials. Metals, 1993, No. 5, pp.152-161.

[149] Gromov V.E., Peregudov O.A., Ivanov Yu.F., Morozov K.V., Alsaraeva K.V. Evolution of the surface layer of rails during long-term operation. Problems of Materials Science, 2015, No. 3, pp. 41-49.

[150] Ivanov Yu. F., Gromov V.E., Yuriev A.A., et al. Gradients of structure and properties of surface layers of differentially hardened rails after long-term operation. Fundamental'nye problemy sovremennogo materialovedenia (Basic Problems of Material Science), 2017, 14(3), pp. 297-305.

[151] Koneva N.A., Kozlov E.V., Trishkina L.I., Lychagin D.V. Long-range stress fields, curvature-torsion of the crystal lattice and stages of plastic deformation. Measurement methods and results. New Methods in Physics and Mechanics of Deformable Solids. In Proceedings of the International Conference. Tomsk: Tomsk State University, 1990, pp. 83-93.

[152] Ivanov Yu.F., Kornet E.V., Kozlov E.V., Gromov V.E. Hardened structural steel: structure and hardening mechanisms. Novokuznetsk: Publishing house of Siberian State Industrial University, 2010, 174p.

[153] Ivanov Yu.F., Gromov V.E., Nikitina E.N. Bainitic structural steel: structure and hardening mechanisms. Novokuznetsk: Publishing house of Siberian State Industrial University, 2015, 177p.

[154] Koneva N., Kiseleva S., Popova N. Evolution of structure and internal stress fields. Austenitic steel. Saarbrucken: LAP LAMBERT Academic Publishing, 2017, 148p.

[155] Yuriev A.A., Gromov V.E., Grishunin V.A., et al. Fracture mechanisms of lamellar pearlite of differentially hardened rails during long-term operation. Fundamental'nye problemy sovremennogo materialovedenia (Basic Problems of Material Science), 2017, 14(4), pp. 438-444.

[156] Ivanov Yu.F., Ivanov Yu.F., Yuriev A.A., Gromov V.E., et al. Transformation of the carbide phase of rails during long-term operation. Bulletin of Higher Educational Institutions. Ferrous Metallurgy, 2018, 61(2), pp. 140-148. https://doi.org/10.17073/0368-0797-2018-2-140-148

[157] Gavrilyuk V.G., Gertsriken D.S., Polushkin Yu.A., Falchenko V.M. The mechanism of cementite decomposition during plastic deformation of steel. Physics of Metals and Metal Science, 1981, 51(1), pp. 147-152.

[158] Gridnev V.N., Gavrilyuk V.G. The decomposition of cementite during plastic deformation of steel. Metal Physics, 1982, 4(3), pp. 74-87.

[159] Mail R.F., Hagel W.C. Advances in the physics of metals. Vol. 3. Moscow: Metallurgy, 1960, pp. 88-156.

[160] Belous M.V., Cherepin V.T. Changes in the carbide phase of steel under the influence of cold plastic deformation. Physics of Metals and Metal Science, 1962, 14(1), pp. 48-54.

[161] Gavrilyuk V.G. Distribution of carbon in steel. Kyiv: Naukova Dumka, 1987, 207p.

[162] Smirnov O.M., Lazarev V.A. Diffusion and redistribution of carbon in iron and its alloys during deformation. Physics of Metals and Metal Science, 1983, 56(1), pp. 115-119.

[163] Bataev A.A. Regularities of plastic deformation of pearlite and the development of effective hardening processes for steels with a heterophase structure. Dissertation of Doctor of Technical Sciences. Novosibirsk, 1995, 398p.

[164] Kozlov E.V., Zakirov D.M., Popova N.A., et al. Substructural-phase transformations during severe plastic deformation of low-carbon ferritic-pearlitic steel. Proceedings of Universities. Physics, 1998, No. 3, pp. 63-71. https://doi.org/10.1007/BF02766420

[165] Gromov V.E., Berdyshev V.A., Kozlov E.V., Petrov V.I., Sarychev V.D., Dorofeev V.V., Ivanov Yu.F., Ignatenko L.N., Popova N.A., Zellermaer V.Ya. Gradient structural-phase states in rail steel. Moscow: Nedra Communications LTD, 2000, 176p.

[166] Guryev A.M., Kozlov E.V., Ignatenko L.N., Popova N.A. Physical foundations of thermocyclic borating of steels. Barnaul: Altai State Technical University, 2000, 177p.

[167] Popova N.A., Zhuleikin S.G., Ignatenko L.N., et al. Formation of gradient structures in pearlitic steel during operation. Bulletin of Tambov University, 2003, 8(4), pp. 589-590.

[168] Veter V.V., Popova N.A., Ignatenko L.N., Kozlov E.V. Fragmented substructure and cracking in low-alloy steel. Bulletin of Higher Educational Institutions. Ferrous Metallurgy, 1994, No. 10, pp. 44-48.

[169] Veter V.V., Zhuleikin S.G., Ignatenko L.N., Kovalenko V.V., Gromov V.E., Popova N.A., Kozlov E.V. Gradient structures arising during plastic deformation of pearlitic steel. Proceedings of Russian Academy of Sciences. Physical Series, 2003, 67(10), pp. 1375-1379.

[170] Gulyaev A.P. Metallurgy. Moscow: Metallurgy, 1978, 647p.

[171] Taran Yu.N. The structure of iron-carbon alloys. Metallurgy and Heat Treatment of Steel. Reference edition. Vol. II. Edited by M.L. Bernstein and A.G. Rakhstadt. Moscow: Metallurgy, 1983, pp.67-83.

[172] Bernstein M. L., Zaimovsky V. A., Kaputkina L. M. Thermomechanical treatment of steel. Moscow: Metallurgy, 1983, 480p.

[173] Kurdyumov V.G., Utevsky L.M., Entin R.I. Transformations in iron and steel. Moscow: Nauka, 1977, 236p.

[174] Romaniv O.N. Fracture toughness of structural steels. Series "Advances in Modern Metal Science". Moscow: Metallurgy, 1979, 176p.

[175] Embrittlement of structural steels and alloys. Edited by K.L. Bryant and S.K. Benergy. Moscow: Metallurgy, 1988, 52p.

[176] Geld P.V., Ryabov R.A., Mokhracheva L.P. Hydrogen and physical properties of metals and alloys. Moscow: Nauka, 1985, 232p.

[177] Grigorovich V.K. Electronic structure and thermodynamics of iron alloys. Moscow: Nauka, 1970, 292p.

[178] Vol A.E. The structure and properties of double metal systems. Moscow: State Publishing House of Physical and Mathematical Literature, 1962, Vol.2, 982p.

[179] Yekh Ya. Heat treatment of steel. Handbook. - Moscow: Metallurgy, 1979, 264p.

[180] Lysak L.I., Nikolin B.I. Physical foundations of heat treatment of steel. Kyiv: Technika. 1975, 304p.

[181] Metallurgy and heat treatment of steel: Handbook. Edited by M.L. Bernstein and A.G. Rakhstadt. Moscow: Metallurgy. Vol. 2, 1983, 386p.

[182] Blanter M.E. Phase transformations during heat treatment of steel. Moscow: Metallurgy, 1962, 268p.

[183] Goodremont E. Special steels. Vols. I and II: Translated from German. Moscow: Metallurgy, 1966, 1274p.

[184] Meskin V.S. Basics of alloying steel. Moscow: Metallurgy, 1964, 684p.

[185] Petrov Yu.N. Defects and diffusionless transformation in steel. Kyiv: Naukova Dumka, 1978, 267p.

[186] Novikov I.I. Theory of heat treatment of metals. Moscow: Metallurgy, 1978, 392p.

[187] Pickering F.B. Physical metallurgy and steel processing. Moscow: Metallurgy, 1982, 184p.

[188] Schastlivtsev V.M., Mirzaev D.A., Yakovleva I.L. Heat treated steel structure. Moscow: Metallurgy, 1994, 288p.

[189] Bernstein M.L., Kaputkina L.M., Prokoshkin S.D. Tempering of steel. Moscow: National Research Technological University "MISiS", 1997, 336p.

[190] Schastlivtsev V.M., Mirzaev D.A., Yakovleva I.L., Okishev K.Yu., Tabatchikova T.I., Khlebnikova Yu.V. Perlite in carbon steels. Yekaterinburg: Ural Branch of the Russian Academy of Sciences, 2006, 312p.

[191] Morozov O.P., Schastlivtsev V.M., Yakovleva I.L. Upper and lower bainite in carbonaceous eutectoid steel. Physics of Metals and Metal Science, 1990, No. 2, pp. 150-159.

[192] Speich G., Swann P.R. Yield strength and transformation substructure of quenched iron-nickel alloys. Journal of Iron and Steel Institute, 1965, 203(4), pp. 480-485.

[193] Belous M.V., Cherepin V.T., Vasiliev M.A. Steel tempering transformations. Moscow: Metallurgy, 1973, 232p.

[194] Belous M.V. Distribution of carbon by states during tempering of hardened steel. Metal Physics. Republican Interdepartmental Collection, 1970, No. 32, pp. 79-82.

[195] Belous M.V., Shatalova L.A., Sheiko Yu.P. The state of carbon in tempered and cold-worked steel. First transformation on tempering. Physics of Metals and Metal Science, 1994, 78(2), pp. 99-106.

[196] Belous M.V., Moskalenok Yu.N., Cherepin V.T., Sheiko Yu.P., Meshashti S. State of carbon in tempered and cold-deformed steel. Volume effects on heating of hardened Fe-C alloys. Physics of Metals and Metal Science, 1995, 80(3), pp. 103-114.

[197] Belous M.V., Novozhilov V.B., Shatalova L.S., Sheiko Yu.P. Distribution of carbon by states in tempered steel. Physics of Metals and Metal Science, 1995. 79(4), pp. 128-137.

[198] Izotov V.I., Kozlova A.G. Distribution of carbon in a package of martensite crystals and its effect on the strength of hardened low-alloy steels. Physics of Metals and Metal Science, 1995, 80(1), pp. 97-111.

[199] Izotov V.I., Filippov G.A. Influence of overcooling at normal γ→α transformation on the distribution of carbon in ferrite of low-alloy steel. Physics of Metals and Metal Science, 1999, 87(4), pp. 72-77.

[200] Speich G.R. Tempering of low-carbon martensite. Transactions of Metallurgical Society of AIME, 1969, 245(10), pp. 2553-2564.

[201] Kalich D., Roberts E.M. On the distribution of carbon in martensite. Transactions of Metallurgical Society of AIME, 1971, 2(10), pp. 2783-2790. https://doi.org/10.1007/BF02813252

[202] Fasiska E.J., Wagenblat H. Dilatation of alpha-iron by carbon. Transactions of Metallurgical Society of AIME, 1967, 239(11), pp. 1818-1820.

[203] Ridley N., Stuart H., Zwell L. Lattice parameters of Fe-C austenite of room temperature. Transactions of Metallurgical Society of AIME, 1969, 246(8), pp. 1834-1836.

[204] Veselov S.I., Spektor E.Z. Dependence of the lattice parameter of austenite on the carbon content at high temperatures. Physics of Metals and Metal Science, 1972, 34(5), pp. 895-896.

[205] Lakhtin Yu.M. Metallurgy and heat treatment of metals. Moscow: Metallurgy, 1977, 407p.

[206] Thomas G., Sarikaya M. Lath martensites in carbon steels - are they bainitic? Proceedings of International Conference on Solid-Solid Phases Transformations, Pittsburgh, Pa, Aug. 10-14, 1981. Warrendale, Pa. 1982, pp. 999-1003.

[207] Sarikaya M., Thomas G., Steeds J.W., et al. Solute element partitioning and austenite stabilization in steels. Proceedings of International Conference on Solid-Solid Phases Transformations, Pittsburgh, Pa, Aug. 10-14, 1981. Warrendale, Pa, 1982, pp. 1421-1425. https://doi.org/10.2172/7031961

[208] Ivanov Yu.F., Popova N.A., Gladyshev S.A., Kozlov E.V. The interaction of carbon with defects and the processes of carbide formation in structural steels. Proceedings "Interaction of Crystal Lattice Defects and Properties." Tula: Tula Polytechnic Institute, 1986, pp. 100-105.

[209] Bhadeshia H.K.D.H. Bainite in steels. Transformation, microstructure and properties. 2nd Ed. The Institute of Materials, London, 2001, 460p.

[210] Schastlivtsev V.M., Kaletina Yu.V., Fokina E.A. Retained austenite in alloy steels. Yekaterinburg: Ural Branch of the Russian Academy of Sciences, 2014, 236p.

[211] Barnard S.J., Smith G.D.W., Saricaya M., Thomas G. Carbon atom distribution in a dual phase steels: atom probe study. Scripta Metallurgica, 1981, 15(4), pp. 387-392. https://doi.org/10.1016/0036-9748(81)90216-7

Structure and Properties of Lengthy Rails after Extreme Long-Term Operation
Materials Research Foundations **106** (2021)

Materials Research Forum LLC
https://doi.org/10.21741/9781644901472

[212] Gromov V.E., Yuriev A.A., Ivanov Yu.F., et al. Redistribution of carbon atoms in differentially hardened rails during long-term operation. Bulletin of Higher Educational Institutions. Ferrous Metallurgy, 2018, 61(6), pp. 56-69. https://doi.org/10.17073/0368-0797-2018-6-454-459

[213] Tushinsky L.I., Bataev A.A., Tikhomirova L.B. Pearlite structure and structural strength of steel. Novosibirsk: VO Nauka, 1993, 280p.

[214] Shtremel M.A. Strength of alloys. Part II: Deformation. Textbook for universities. Moscow: National Research Technological University "MISiS", 1997, 527p.

[215] Belenkiy B.Z., Farber B.M., Goldstein M.I. Estimates of the strength of low-carbon low-alloy steels from structural data. Physics of Metals and Metal Science, 1975, 39(3), pp. 403-409.

[216] Prnka T. Quantitative relations between the parameters of dispersed precipitates and mechanical properties of steels. Metallurgy and Heat Treatment of Steel, 1979, No. 7, pp. 3-8.

[217] Trefilov V.I., Moiseev V.I., Pechkovsky E.P., Gornaya I.D., Vasiliev A.D. Strain hardening and fracture of polycrystalline metals. Kyiv: Naukova Dumka, 1987, 248p.

[218] Static strength and fracture mechanics of steels: Collection of research papers. Translated from German. Edited by V. Dahl and V. Anton. Moscow: Metallurgy, 1986, 566p.

[219] Shur E.A. Damage to rails. Moscow: Intekst, 2012, 192p.

[220] Sheinman E. Wear of rails. A review of the American press. Journal of Friction and Wear, 2012, 33(4), pp. 308-314. https://doi.org/10.3103/S1068366612040101

[221] Gromov V.E., Kozlov E.V., Bazaykin V.I., Zellermaer V.Ya., Ivanov Yu.F., et al. Physics and mechanics of drawing and volumetric stamping. Moscow: Nedra, 1997, 293p.

[222] Koneva N.A., Lychagin D.V., Teplyakova L.A., Kozlov E.V. Dislocation-disclination substructures and hardening. Theoretical and Experimental Study of Disclinations. Leningrad: Institute of Physics and Technology, 1984, pp. 116-126.

[223] McLean D. Mechanical properties of metals. Translated from English. Moscow: Metallurgy, 1965, 431p.

[224] Predvoditelev A.A. The current state of research on dislocation ensembles. In the book: Problems of modern crystallography. Moscow: Nauka, 1975, pp. 262-275.

[225] Embyri I.D. Strengthening by dislocations structure. Strengthening Method in Crystals. Applied Science Publishers, 1971, pp. 331-402.

[226] Mott N.F., Nabarro F.R.N. The distribution of dislocations in slip band. Proceedings of Physical Society, 1940, 52(1), pp. 86-93. https://doi.org/10.1088/0959-5309/52/1/312

[227] Gromov V.E., Yuriev A.A., Ivanov Yu.F., et al. Analysis of the mechanisms of deformation hardening of rail steel during long-term operation. Problems of Ferrous Metallurgy and Material Science, 2017, No. 3. pp. 76-84.

[228] Tyumentsev A.N., Korotaev A.D., Ditenberg I.A., Pinzhin Yu.P., Chernov V.M. Regularities of plastic deformation in high-strength and nanocrystalline metallic materials. Novosibirsk: Siberian Branch of Russian Academy of Sciences: Science: Publishing house SB RAS, 2018, 256p.

[229] Thomas G., Goringe M.J. Transmission electron microscopy of materials. Moscow: Nauka, 1983, 320p.

[230] Gorelik S.S., Skakov Yu. A., Rastorguev L.N. X-ray and electron-optical analysis. Textbook for universities. 3rd ed. National Research Technological University "MISiS", 1994, 328p.

[231] Kormyshev V.E., Gromov V.E., Ivanov Yu.F., Glezer A.M., Yuriev A.A., Semin A.P., Sundeev R.V. Structural phase states and properties of rails after long-term operation. Materials Letters, 2020, 268, 127499, 4p. https://doi.org/10.1016/j.matlet.2020.127499

[232] Kormyshev V.E., Ivanov Yu.F., Gromov V.E., Yuriev A.A., Polevoy E.V. Structure and properties of differentially hardened 100-m rails after extremely long-term operation. Fundamental'nye problemy sovremennogo materialovedenia (Basic Problems of Material Science), 2019, 16(4), pp. 538-546.

[233] Kormyshev V.E., Polevoy E.V., Yuriev A.A., Gromov V.E., Ivanov Yu.F. Formation of the structure of differentially hardened 100-meter rails during long-term operation. Bulletin of Higher Educational Institutions. Ferrous Metallurgy, 2020, 63(2), pp. 108-115. https://doi.org/10.17073/0368-0797-2020-2-108-115

[234] Kormyshev V.E., Ivanov Yu.F., Yuriev A.A., Polevoy E.V., Gromov V.E., Glezer A.M. Evolution of structural-phase states and properties of differentially hardened 100-meter rails during extremely long-term operation. Communication 1. Structure and properties of rail steel before the operation. Problems of Ferrous Metallurgy and Material Science, 2019, No. 4, pp. 50-56.

[235] Kormyshev V.E., Gromov V.E., Ivanov Yu.F., Glezer A.M. Structure of differentially hardened rails under severe plastic deformation. Deformation and Destruction of Materials, 2020, No. 8, pp. 16-20.

[236] Kormyshev V.E., Yuryev A.A., Gromov V.E., Ivanov Yu.F., Rubannikova Yu.A., Polevoy E.V. Stages of the transformation of lamellar pearlite of differentially hardened rails during long-term operation. Problems of Ferrous Metallurgy and Material Science, 2020, No. 2, pp. 1-4.

[237] Yao M.J., Welsch E., Ponge D., Haghighat S.M.H., Sandlöbes S., Choi P., Herbig M., Bleskov I., Hickel T., Lipinska-Chwalek M., Shantraj P., Scheu C., Zaefferer S., Gault B., Raabe

D. Strengthening and strain hardening mechanisms in a precipitation-hardened high-Mn lightweight steel. Acta Materia, 2017, 140, pp. 258-273. https://doi.org/10.1016/j.actamat.2017.08.049

[238] Friedman L.H., Chrzan D.C. Scaling theory of the hall-petch relation for multilayers. Physical Review Letters, 1998, 81, p. 2715. https://doi.org/10.1103/PhysRevLett.81.2715

[239] Han Y., Shi J., Xu L., Cao W.Q., Dong H. TiC precipitation induced effect on microstructure and mechanical properties in low carbon medium manganese steel. Materials Science and Engineering A, 2011, 530, pp. 643-651. https://doi.org/10.1016/j.msea.2011.10.037

[240] Zurob H.S., Hutchinson C.R., Brechet Y., Purdy G. Modeling recrystallization of microalloyed austenite: effect of coupling recovery, precipitation and recrystallization. Acta Materia, 2002, 50, pp. 3077-3094. https://doi.org/10.1016/S1359-6454(02)00097-6

[241] Shima Y., Ishikawa Y., Nitta H., Yamazaki Y., Mimura K., Isshiki M., Iijima Y. Self-diffusion along dislocations in ultra high purity iron. Material Transactions, 2002, 43, p. 173. https://doi.org/10.2320/matertrans.43.173

[242] Morito S., Nishikawa J., Maki T. Dislocation density within lath martensite in Fe-C and Fe-Ni alloys. ISIJ International, 2003, 43, pp. 1475-1477. https://doi.org/10.2355/isijinternational.43.1475

[243] Huthcinson B., Hagstrom J., Karlsson O., Lindell D., Tornberg M., Lindberg F., Thuvander M. Microstructures and hardness of as-quenched martensites (0.1-0.5%C). Acta Materia, 2011, 59, pp. 5845-5858. https://doi.org/10.1016/j.actamat.2011.05.061

[244] Kim J.G., Enikeev N.A, Seol J.B., Abramova M.M., Karavaeva M.V., Valiev R.Z., Park C.G., Kim H. S. Superior strength and multiple strengthening mechanisms in nanocrystalline TWIP steel. Scientific Reports, 2018, 8, p. 11200. https://doi.org/10.1038/s41598-018-29632-y

[245] Sevillano J.G. An alternative model for the strain hardening of FCC alloys that twin, validated for twinning-induced plasticity steel. Scripta Materialia, 2009, 60, pp. 336-339. https://doi.org/10.1016/j.scriptamat.2008.10.035

[246] Bouaziz O., Allain S., Scott S. Effect of grain and twin boundaries on the hardening mechanisms of twinning-induced plasticity steels. Scripta Materialia, 2008, 58, pp. 484-487. https://doi.org/10.1016/j.scriptamat.2007.10.050

[247] Senkov O.N., Scott J.M., Senkova S.V., Miracle D.B., Woodward C.F. Microstructure and room temperature properties of a high-entropy TaNbHfZrTi alloy. Journal of Alloys and Compounds, 2011, 509, pp. 6043-6048. https://doi.org/10.1016/j.jallcom.2011.02.171

[248] Ganji R.S., Karthik P.S., Rao K.B.S., Rajulapati K.V. Strengthening mechanisms in equiatomic ultrafine grained AlCoCrCuFeNi high-entropy alloy studied by micro- and

nanoindentation methods. Acta Materia, 2017, 125, pp. 58-68.
https://doi.org/10.1016/j.actamat.2016.11.046

[249] Silva R.A., Pinto A.L., Kuznetsov A., Bott I.S. Precipitation and grain size effects on the tensile strain-hardening exponents of an API X80 steel pipe after high-frequency hot-induction bending. Metals, 2018, 8, pp. 168-180. https://doi.org/10.3390/met8030168

[250] Hosford W.F. Mechanical Behavior of Materials, 2nd ed. Cambridge University Press: Cambridge, UK, 2010, 163p.

[251] Morales E.V., Gallego J., Kestenbachz H.-J. On coherent carbonitride precipitation in commercial microalloyed steels. Philosophical Magazine Letters, 2003, 83, pp. 79-87. https://doi.org/10.1080/0950083021000056632

[252] Morales E.V., Galeano Alvarez N.J., Morales A.M., Bott I.S. Precipitation kinetics and their effects on age hardening in a Fe-Mn-Si-Ti martensitic alloy. Materials Science and Engineering: A, 2012, 534, pp. 176-185 https://doi.org/10.1016/j.msea.2011.11.056

[253] Sieurin H., Zander J., Sandström R. Modelling solid solution hardening in stainless steels. Materials Science and Engineering: A, 2006, 415, pp. 66-71. https://doi.org/10.1016/j.msea.2005.09.031

[254] Chatterjee S., Wang H.S., Yang J.R., Bhadeshia H.K.D.H. Mechanical stabilization of austenite. Materials Science and Technology, 2006, 22, pp. 641-644. https://doi.org/10.1179/174328406X86128

[255] Fine M.E., Isheim D. Origin of copper precipitation strengthening in steel revisited. Scripta Materialia, 2005, 53, pp.115-118. https://doi.org/10.1016/j.scriptamat.2005.02.034

[256] Goldstein M.I., Farber B.M. Dispersion hardening of steel. Moscow: Metallurgy, 1979, 208p.

[257] Ivanisenko Yu., Maclaren I., Souvage X., Valiev R. Z., Fecht H. J. Shear-induced α→γ transformation in nanoscale Fe-C composite. Acta Materia, 2006, 54, pp. 1659-1669. https://doi.org/10.1016/j.actamat.2005.11.034

[258] Maisuradze M., Ryzhkov M., Lebedev D. Microstructure and mechanical properties of martensitic high-strength engineering steel. Metallurgist, 2020, 64, pp. 640-651. https://doi.org/10.1007/s11015-020-01040-6

[259] Terentyev V.F. Model of the physical limit of fatigue of metals and alloys. Reports of the USSR Academy of Sciences, 1969, 185(2), pp. 325-326.

[260] Rybin V.V. Large plastic deformation and destruction of metals. Moscow: Metallurgy, 1986, 224p.

[261] Rybin V.V., Vergazov A.N., Likhachev V.A. Viscous destruction of molybdenum as a consequence of structure fragmentation. Physics of Metals and Metal Science, 1974, 37(3), pp. 620-624.

Structure and Properties of Lengthy Rails after Extreme Long-Term Operation
Materials Research Foundations **106** (2021)

Materials Research Forum LLC
https://doi.org/10.21741/9781644901472

[262] Fatigue of steels modified with high-intensity electron beams. Edited by V.E. Gromova, Yu.F. Ivanova. Novokuznetsk: Publishing house "Inter-Kuzbass", 2012, 403p.

[263] Kormyshev V.E., Polevoi E.V., Yur'ev A.A., Gromov V.E., and Ivanov Yu.F. The structural formation in differentially-hardened 100-meter-long rails during long-term operation. Steel in Translation, 2020, 50(2), pp. 77-83. https://doi.org/10.3103/S0967091220020047

[264] Kormyshev V.E., Ivanov Yu.F., Gromov V.E., Yuriev A.A., Rubannikova Yu.A., Semin A.P. Formation of the fine surface of long rails on differentiated hardening. Journal of Surface Investigation: X-ray, Synchrotron and Neutron Techniques, 2020, 14(6), pp. 1186-1189. https://doi.org/10.1134/S1027451020060099

Structure and Properties of Lengthy Rails after Extreme Long-Term Operation
Materials Research Foundations **106** (2021)

Materials Research Forum LLC
https://doi.org/10.21741/9781644901472